BASICS OF MOLECULAR PHYSICS

BASICS OF MOLECULAR PHYSICS

Rajnish Pandit

ANMOL PUBLICATIONS PVT. LTD.
NEW DELHI - 110 002 (INDIA)

ANMOL PUBLICATIONS PVT. LTD.
H.O.: 4374/4B, Ansari Road, Daryaganj,
New Delhi-110 002 (India)
Ph.: 23278000, 23261597
B.O.: No. 1015, Ist Main Road, BSK IIIrd Stage
IIIrd Phase, IIIrd Block,
Bangalore - 560 085 (India)
Visit us at: www.anmolpublications.com

Basics of Molecular Physics

First Published, 2008
ISBN 978-81-261-3567-7

PRINTED IN INDIA

Printed at Mehra Offset Press, Delhi.

Contents

Preface

This book provides a comprehensive and up-to-date account of the field of low energy positrons and positronium within atomic and molecular physics. It begins with an introduction to the field, discussing the background to low energy positron beams, and then covers topics such as total scattering cross sections, elastic scattering, positronium formation, excitation and ionization, annihilation and positronium interactions. Each chapter contains a blend of theory and experiment, giving a balanced treatment of all the topics. The book will be useful for graduate students and researchers in physics and chemistry. It is ideal for those wishing to gain rapid, in-depth knowledge of this unique branch of molecular physics.

Author

Chapter 1

The Density Matrix

PURE STATES AND MIXED STATES

Our treatment here more or less follows that of Sakurai, beginning with two imagined Stern-Gerlach experiments. In that experiment, a stream of (non-ionized) silver atoms from an oven is directed through an inhomogeneous vertical magnetic field, and the stream splits into two. The silver atoms have nonzero magnetic moments, and a magnetic moment in an inhomogeneous magnetic field experiences a nonzero force, causing the atom to veer from its straight line path, the magnitude of the deflection being proportional to the component of the atom's magnetic moment in the vertical (field) direction. The observation of the beam splitting into two, and no more, means that the vertical component of the magnetic moment, and therefore the associated angular momentum, can only have *two* different values. From the basic analysis of rotation operators and the properties of angular momentum that follow, this observation forces us to the conclusion that the total angular momentum of a silver atom is $\frac{1}{2}\hbar$. Ordinary orbital angular momenta cannot have half-integer values; this experiment was one of the first indications that the electron has a spin degree of freedom, an angular momentum that cannot be interpreted as orbital angular momentum of constituent parts. The silver atom has 47 electrons, 46 of them have total spin and orbital momenta that separately cancel, the 47th has no orbital angular momentum, and its spin is the entire angular momentum of the atom.

Here we shall use the Stern-Gerlach stream as an example of a large collection of quantum systems (the atoms) to clarify just how to describe such a collection, often called an *ensemble*. To avoid unnecessary complications, we only consider the *spin* degrees of freedom. We begin by examining two different streams:

Suppose experimentalist *A* prepares a stream of silver atoms such that each atom is in the spin state ψ_A:

$$|\psi_A\rangle = \frac{1}{\sqrt{2}}(|\uparrow\rangle + |\downarrow\rangle)$$

Meanwhile, experimentalist *B* prepares a stream of silver atoms which is a *mixture*: half the atoms are in state $|\uparrow\rangle$ and half are in the state$|\downarrow\rangle$: call this mix *B*.

Question: can we distinguish the *A* stream from the *B* stream?

Evidently, not by measuring the spin in the *z*-direction! Both will give up 50per cent of the time, down 50per cent. But: we *can* distinguish them by measuring the spin in the *x*-direction: the ψ_A quantum state is in fact just that of a spin in the *x*-direction, so it will give "up" in the *x*-direction every time from now on we call it$|\uparrow_x\rangle$, whereas the state $|\uparrow\rangle$ ("up" in the *z*-direction) will yield "up" in the *x*-direction only 50per cent of the time, as will$|\downarrow\rangle$.

The state $\psi_A = |\uparrow_x\rangle$ is called a *pure* state, it's the kind of quantum state we've been studying this whole course.

The stream *B*, in contrast, is in a *mixed* state: the kind that actually occurs to a greater or lesser extent in a real life stream of atoms, different pure quantum states occurring with different probabilities, but with no phase coherence between them. In other words, these relative probabilities in *B* of different quantum states do *not* derive from probability amplitudes, as they do in finding the probability of spin up in stream *A*: the probabilities of the different quantum states in the mixed state *B* are exactly like classical probabilities.

That being said, though, to find the probability of measuring spin up in some such mixed state, one *first* uses the classical-type probability for each component state, *then* for each quantum state in the mix, one finds the probability of spin up *in that state* by the standard quantum technique.

Theerefore, for a mixed state in which the system is in state $|\psi_i\rangle$ with probability w_i, $\Sigma w_i = 1$ the expectation value of an operator A is

$$\langle A\rangle = \sum w_i \langle\psi_i|A|\psi_i\rangle$$

and we should emphasize that these $|\psi_i\rangle$ do *not* need to be orthogonal (but they are of course normalized): for example one could be $|\uparrow_x\rangle$, another $|\uparrow_z\rangle$. (We put the usually omitted z in for emphasis.)

THE DENSITY MATRIX

The equation for the expectation value <A> can be written:

$$\langle A\rangle = \text{Trace}(\hat{\rho}A) \quad \text{where } \hat{\rho} = \sum w_i|\psi_i\rangle\langle\psi_i|.$$

To see exactly how this comes about, recall that for an operator B in a finite-dimensional vector space with an orthonormal basis set, $TrB = \sum_{j=1}^{n}\langle j|B|j\rangle = B_{jj}$, where the repeated suffix implies summation of the diagonal matrix elements of the operator. Therefore,

$$\begin{aligned}\text{Tr}(\hat{\rho}A) &= \sum_{j=1}^{n}\sum_{i=1}^{n}\langle j|w_i|\psi_i\rangle\langle\psi_i|A|j\rangle \\ &= \sum_{j=1}^{n}\sum_{i=1}^{n}\langle\psi_i|A|j\rangle\langle j|w_i|\psi_i\rangle \\ &= \sum_{i=1}^{n} w_i\langle\psi_i|A|\psi_i\rangle\end{aligned}$$

since $\Sigma|\, j \rangle\langle j\,| = 1$, the identity. This $\hat{\rho}$ is called the *density matrix*: its matrix form is made explicit by considering states $|\,\psi_i\rangle$ in a finite N-dimensional vector space (such as spins or angular momenta) $|\,\psi_i\rangle = \sum_j (v_i)_j\,|\,j\rangle$ where the $|\,j\rangle$ are an orthonormal basis set, and $(v_i)_j$ is the jth component of a normalized vector V_i. It is convenient to express $\hat{\rho}$ in terms of kets and bras belonging to this orthonormal basis,

$$\hat{\rho} = \sum w_i |\psi_i\rangle\langle\psi_i| = \sum_{i,j,k} w_i (V_i)_j (V_i^\dagger)_k |j\rangle\langle k| = \sum_{j,k} \rho_{jk} |j\rangle\langle k|$$

and evidently

$$\langle A\rangle = \text{Trace}(\hat{\rho}A) = \sum_{n,j,k} \langle n|\rho_{jk}|j\rangle\langle k|A|n\rangle = \sum_{j,k} \rho_{jk} \langle k|A|j\rangle = \sum_{j,k} \rho_{jk} A_{kj}.$$

Since ρ_{jk} is just a *number*, $\langle n|\rho_{jk}|j\rangle = \rho_{jk}\langle n|j\rangle = \rho_{jk}\delta_{nj}$.

$\text{Trace}(\hat{\rho}A)$ is *basis-independent*, the trace of a matrix being unchanged by a unitary transformation, since it follows from $\text{Tr}ABC = \text{Tr}BCA$ that $\text{Tr}\,U^\dagger AU = \text{Tr}\,AUU^\dagger = \text{Tr}\,A$ for $UU^\dagger = 1$.

Note that since the vectors V_i are normalized, $\sum_j (V_i)_j (V_i^\dagger)_j = 1$, with the i not summed over, and $\sum w_i = 1$, it follows that

$$Tr\hat{\rho} = 1$$

(also evident by putting $A = 1$ in the equation for $\langle A\rangle$).

For a system in a *pure* quantum state $|\psi\rangle$, $\hat{\rho} = |\psi\rangle\langle\psi|$, just the projection operator into that state, and

$$\hat{\rho}^2 = \hat{\rho},$$

as for all projection operators.

It's worth spelling out how this differs from the mixed state by looking at the form of the density matrix.

For the pure state $|\psi\rangle$, if a basis is chosen so that $|\psi\rangle$ is a member of the basis (this can always be done), $\hat{\rho}$ is a matrix with every element zero except the one diagonal element corresponding to $|\psi\rangle\langle\psi|$, which will be unity. Obviously, $\hat{\rho}^2 = \hat{\rho}$. This is less obvious in a general basis, where will not necessarily be diagonal. But the statement remains true under a transformation to a new basis.

For a mixed state, let's say for example a mixture of orthogonal states $|\psi_1\rangle, |\psi_2\rangle$, if we choose a basis including both states, the density matrix will be diagonal with just two entries w_1, w_2 Both these numbers must be less than unity, so $\hat{\rho}^2 \neq \hat{\rho}$. A mix of nonorthogonal states is left as an exercise for the reader.

SOME SIMPLE EXAMPLES

First, our case *A* above (pure state): all spins in state.

$$|\uparrow_x\rangle = (1/\sqrt{2})(|\uparrow\rangle + |\downarrow\rangle)$$

In the standard $|\uparrow\rangle, |\downarrow\rangle$ basis,

$$\hat{\rho} = |\uparrow_x\rangle\langle\uparrow_x| = \begin{pmatrix} 1/\sqrt{2} \\ 1/\sqrt{2} \end{pmatrix} \begin{pmatrix} 1/\sqrt{2} & 1/\sqrt{2} \end{pmatrix} = \begin{pmatrix} 1/2 & 1/2 \\ 1/2 & 1/2 \end{pmatrix}$$

and

$$\langle s_x \rangle = \mathrm{Tr}(\hat{\rho} s_x) = \frac{\hbar}{2} \mathrm{Tr} \begin{pmatrix} 1/2 & 1/2 \\ 1/2 & 1/2 \end{pmatrix} \begin{pmatrix} 0 & 1 \\ 1 & 0 \end{pmatrix} = \frac{\hbar}{2}$$

$$\langle s_z \rangle = \mathrm{Tr}(\hat{\rho} s_z) = \frac{\hbar}{2} \mathrm{Tr} \begin{pmatrix} 1/2 & 1/2 \\ 1/2 & 1/2 \end{pmatrix} \begin{pmatrix} 1 & 0 \\ 0 & -1 \end{pmatrix} = 0.$$

Notice that $\hat{\rho}^2 = \hat{\rho}$.

Now, case *B* (50-50 mixed up and down):

50per cent in the state $|\uparrow\rangle$, 50per cent $|\downarrow\rangle$.

The density matrix is

$$\hat{\rho} = \frac{1}{2}|\uparrow\rangle\langle\uparrow| + \frac{1}{2}|\downarrow\rangle\langle\downarrow|$$

$$= \frac{1}{2}\begin{pmatrix} 1 \\ 0 \end{pmatrix}\begin{pmatrix} 1 & 0 \end{pmatrix} + \frac{1}{2}\begin{pmatrix} 0 \\ 1 \end{pmatrix}\begin{pmatrix} 0 & 1 \end{pmatrix} = \frac{1}{2}\begin{pmatrix} 1 & 0 \\ 0 & 1 \end{pmatrix}.$$

This is proportional to the unit matrix, so

$$\mathrm{Tr}\, \hat{\rho} s_x = \frac{1}{2}\frac{\hbar}{2} \mathrm{Tr}\, \sigma_x = 0,$$

and similarly for s_y and s_z, since the Pauli *s*-matrices are all traceless. Note also that $\hat{\rho}^2 = \frac{1}{2}\hat{\rho} \neq \hat{\rho}$, as is true for all mixed states.

Finally, a 50-50 mixed state relative to the x-axis:

That is, 50per cent of the spins in the state $|\uparrow_x\rangle = (1/\sqrt{2})(|\uparrow\rangle + |\downarrow\rangle)$, "up" along the x-axis, and 50per cent in $|\downarrow_x\rangle = (1/\sqrt{2})(|\uparrow\rangle - |\downarrow\rangle)$, "down" in the x-direction.

It is easy to check that

$$\hat{\rho} = \frac{1}{2}|\uparrow_x\rangle\langle\uparrow_x| + \frac{1}{2}|\downarrow_x\rangle\langle\downarrow_x| = \frac{1}{2}\begin{pmatrix} 1/2 & 1/2 \\ 1/2 & 1/2 \end{pmatrix} + \frac{1}{2}\begin{pmatrix} 1/2 & -1/2 \\ -1/2 & 1/2 \end{pmatrix} = \frac{1}{2}\begin{pmatrix} 1 & 0 \\ 0 & 1 \end{pmatrix}$$

This is exactly the same density matrix we found for 50per cent in the state $|\uparrow\rangle$, 50per cent $|\downarrow\rangle$!

The reason is that both formulations describe a state about which we know nothing we are in a state of *total ignorance*, the spins are completely random, all directions are equally likely. The density matrix describing such a state cannot depend on the direction we choose for our axes.

Another two-state quantum system that can be analysed in the same way is the polarization state of a beam of light, the basis states being polarization in the x-direction and polarization in the y-direction, for a beam traveling parallel to the z-axis. Ordinary unpolarized light corresponds to the random mixed state, with the same density matrix as in the last example above.

TIME EVOLUTION OF THE DENSITY MATRIX

In the mixed state, the quantum states evolve independently according to Schrödinger's equation, so

$$i\hbar\frac{d\hat{\rho}}{dt} = \sum w_i H\,|\psi_i\rangle\langle\psi_i| - \sum w_i\,|\psi_i\rangle\langle\psi_i|\,H = [H, \hat{\rho}].$$

Note that this has the opposite sign from the evolution of a Heisenberg operator, not surprising since the density operator is made up of Schrödinger bras and kets.

The equation is the quantum analogue of *Liouville's theorem* in statistical mechanics. Liouville's theorem describes the evolution in time of an ensemble of identical classical systems, such as many boxes each filled with the same amount of the same gas at the same temperature, but the positions and momenta of the individual atoms are randomly different in each. Each box can be classically described by a single point in a huge dimensional space, a space having six dimensions for each atom (position and momentum, we ignore possible internal degrees of freedom). The whole ensemble, then, is a gas of these points in this huge space, and the rate of change of local density of this gas, from Hamilton's equations, is $\partial\rho/\partial t = -\{\rho, H\}$, the bracket now being a Poisson bracket. Anyway, this is the classical precursor of, and the reason for the name of, the density matrix.

THERMAL EQUILIBRIUM

A system in thermal equilibrium is represented in statistical mechanics by a *canonical ensemble.* If the eigenstate $|i\rangle$ of the Hamiltonian has energy E_i, the relative probability of the system being in that state is $e^{-E_i/kT} = e^{-\beta E_i}$ in the standard notation. Therefore the density matrix is:

$$\hat{\rho} = \frac{1}{z}\sum_i e^{-\beta E_i} |i\rangle\langle i| = \frac{e^{-\beta H}}{Z},$$

where

$$Z = \sum_i e^{-\beta E_i} = \mathrm{Tr}\, e^{-\beta H}$$

Notice that in this formulation, apart from the normalization constant Z, the density operator is analogous to the propagator $U(t) = e^{-iHt/\hbar}$ for an imaginary time $t = -i\hbar\beta$. Incidentally, for interacting quantum fields, the propagator can be constructed as a set of Feynman diagrams corresponding to all possible sequences of particle scatterings by interaction. To find the thermodynamic properties of a field theory at finite temperature, essentially the same set of diagrams is used to find the free energy: the diagrams now describe the system propagating for a finite imaginary time, the same mathematical tools can be used.

At zero temperature ($\beta = \infty$) the probability coefficients $w_i = e^{-\beta E_i / z}$ are all zero except for the ground state: the system is in a pure state, and the density matrix has every element zero except for a single element on the diagonal. At infinite temperature, all the w_i are equal: the density matrix is just $1/N$ times the unit matrix, where N is the total number of states available to the system. In fact, the *entropy* of the system can be expressed in terms of the density matrix: $S = -kTr(\hat{\rho} In \hat{p})$. This is not as bad as it looks: both operators are diagonal in the energy subspace.

Chapter 2

Charged Particle in a Magnetic Field

INTRODUCTION

Classically, the force on a charged particle in electric and magnetic fields is given by the Lorentz force law:

$$\vec{F} = q\left(\vec{E} + \frac{\vec{v} \times \vec{B}}{c}\right)$$

This velocity-dependent force is quite different from the conservative forces from potentials that we have dealt with so far, and the recipe for going from classical to quantum mechanics replacing momenta with the appropriate derivative operators has to be carried out with more care. We begin by demonstrating how the Lorentz force law arises classically in the Lagrangian and Hamiltonian formulations.

LAWS OF CLASSICAL MECHANICS

That the Principle of Least Action leads to the Euler-Lagrange equations for the Lagrangian L:

$$\frac{d}{dt}\left(\frac{\partial L(q_i, \dot{q}_i)}{\partial \dot{q}_i}\right) - \frac{\partial L(q_i, \dot{q}_i)}{\partial q_i} = 0, \ q_i, \dot{q}_i \text{ being coordinates and velocities.}$$

The canonical momentum p_i is defined by the equation

$$p_i = \frac{\partial L}{\partial \dot{q}_i}$$

and the Hamiltonian is defined by performing a Legendre transformation of the Lagrangian:

$$H(q_i, p_i) = \Sigma p_i q_i - L(q_i, \overline{q}_i) - L(q_i, \overline{q}_i)$$

It is straightforward to check that the equations of motion can be written:

$$\overline{q}_i = \frac{\partial H}{\partial p_i}, \overline{p}_i = -\frac{\partial H}{\partial q_i}$$

These are known as *Hamilton's Equations.* Note that if the Hamiltonian is independent of a particular coordinate q_i, the corresponding momentum p_i remains constant. (Such a coordinate is termed *cyclic,* because the most common example is an angular coordinate in a spherically symmetric Hamiltonian, where angular momentum remains constant.)

For the conservative forces we have been considering so far, $L = T - V$, $H = T + V$, with T the kinetic energy, V the potential energy.

POISSON BRACKETS

Any dynamical variable f in the system is some function of the q_i's and p_i's and (assuming it does not depend explicitly on time) its development is given by:

$$\frac{d}{dt} f(q_i, p_i) = \frac{\partial f}{\partial q_i} \dot{q}_i + \frac{\partial f}{\partial p_i} \dot{p}_i = \frac{\partial f}{\partial q_i}\frac{\partial H}{\partial p_i} - \frac{\partial f}{\partial p_i}\frac{\partial H}{\partial q_i} = \{f, H\}.$$

The curly brackets are called *Poisson Brackets,* and are defined for any dynamical variables as:

$$\{A, B\} = \frac{\partial A}{\partial q_i}\frac{\partial B}{\partial p_i} - \frac{\partial A}{\partial p_i}\frac{\partial B}{\partial q_i}.$$

We have shown from Hamilton's equations that for any variable $\overline{f} = (f, H)$.

It is easy to check that for the coordinates and canonical momenta,

$$\{q_i, q_j\} = 0 = \{p_i, p_j\}, \quad \{q_i, p_j\} = \delta_{ij}.$$

This was the classical mathematical structure that led Dirac to link up classical and quantum mechanics: he realised that the Poisson brackets were the classical version of the commutators, so a classical canonical momentum must correspond to the quantum differential operator in the corresponding coordinate.

PARTICLE IN A MAGNETIC FIELD

The Lorentz force is velocity dependent, so cannot be just the gradient of some potential. Nevertheless, the classical particle path is still given by the Principle of Least Action. The electric and magnetic fields can be written in terms of a scalar and a vector potential:

$$\vec{B} = \vec{\nabla} \times \vec{A}, \quad \vec{E} = -\vec{\nabla}\varphi - \frac{1}{c}\frac{\partial \vec{A}}{\partial t}.$$

The right Lagrangian turns out to be:

$$L = \tfrac{1}{2} m\vec{v}^2 - q\varphi + \frac{q}{c}\vec{v}.\vec{A}$$

(*Note*: if you're familiar with Relativity, the interaction term here looks less arbitrary: the relativistic version would have the relativistically invariant $(q/c)\int A^{\mu} dx_{\mu}$ added to the action integral, where the four-potential $A_{\mu} = (A, \varphi)$ and $dx_{\mu} = (dx_1, dx_2, dx_3, cdt)$. This is the simplest possible invariant interaction between the electromagnetic field and the particle's four-velocity. Then in the nonrelativistic limit, $(q/c)\int A^{\mu} dx_{\mu}$ just becomes $\int q(\vec{v}.\vec{A}/c - \varphi)dt$.)

Note that for zero vector potential, the Lagrangian has the usual T - V form.

For this one-particle problem, the general coordinates q_i are just the Cartesian co-ordinates $x_i = (x_1, x_2, x_3)$, the position of the particle, and the are the three components $\dot{x}_i = v_i$ of the particle's velocity.

The important *new* point is that the canonical momentum

$$p_i = \frac{\partial L}{\partial \dot{q}_i} = \frac{\partial L}{\partial \dot{x}_i} = mv_i + \frac{q}{c}A_i$$

is no longer mass × velocity there is an extra term!

The Hamiltonian is

$$\begin{aligned} H(q_i, p_i) &= \sum p_i \dot{q}_i - L(q_i, \dot{q}_i) \\ &= \sum \left(mv_i + \frac{q}{c}A_i \right) v_i - \tfrac{1}{2} m\vec{v}^2 + q\varphi - \frac{q}{c}\vec{v}.\vec{A} \\ &= \tfrac{1}{2} m\vec{v}^2 + q\varphi \end{aligned}$$

Reassuringly, the Hamiltonian just has the familiar form of kinetic energy plus potential energy. However, to get Hamilton's equations of motion, the Hamiltonian has to be expressed solely in terms of the coordinates and canonical momenta. That is,

$$H = \frac{\left(\vec{p} - q\vec{A}(\vec{x}, t)/c\right)^2}{2m} + q\varphi(\vec{x}, t)$$

where we have noted explicitly that the potentials mean those at the position $\vec{x}$ of the particle at time t.

Let us now consider Hamilton's equations

$$\dot{x}_i = \frac{\partial H}{\partial p_i}, \quad \dot{p}_i = -\frac{\partial H}{\partial x_i}$$

It is easy to see how the first equation comes out, bearing in mind that

$$p_i = mv_i + \frac{q}{c}A_i = m\dot{x}_i + \frac{q}{c}A_i.$$

The second equation yields the Lorentz force law, but is a little more tricky. The first point to bear in mind is that dp/dt is *not* the acceleration, the A term also varies in time, and in a quite complicated way, since it is *the field at a point moving with the particle.*

That is,

$$\dot{p}_i = m\ddot{x}_i + \frac{q}{c}\dot{A}_i = m\ddot{x}_i + \frac{q}{c}\left(\frac{\partial A_i}{\partial t} + v_j \nabla_j A_i\right).$$

The right-hand side of the second Hamilton equation $\overline{p}_i = -\frac{\partial H}{\partial x_i}$ is

$$-\frac{\partial H}{\partial x_i} = \frac{\left(\vec{p} - q\vec{A}(\vec{x},t)/c\right)}{m} \cdot \frac{q}{c} \cdot \frac{\partial \vec{A}}{\partial x_i} - q\frac{\partial \varphi(\vec{x},t)}{\partial x_i}$$

$$= \frac{q}{c} v_j \nabla_i A_j - q\nabla_i \varphi.$$

Putting the two sides together, the Hamilton equation reads:

$$m\ddot{x}_i = -\frac{q}{c}\left(\frac{\partial A_i}{\partial t} + v_j \nabla_j A_i\right) + \frac{q}{c} v_j \nabla_i A_j - q\nabla_i \varphi.$$

Using $\vec{v} \times (\vec{\nabla} \times \vec{A}) = \vec{\nabla}(\vec{v}.\vec{A}) - (\vec{v}.\vec{\nabla})\vec{A}$, $\vec{B} = \vec{\nabla} \times \vec{A}$, and the expressions for the electric and magnetic fields in terms of the potentials, the Lorentz force law emerges:

$$m\ddot{\vec{x}} = q\left(\vec{E} + \frac{\vec{v} \times \vec{B}}{c}\right)$$

QUANTUM MECHANICS OF A PARTICLE IN A MAGNETIC FIELD

We make the standard substitution:

$\vec{p} = -i\hbar\vec{\nabla}$, so that $[x_i, p_j] = i\hbar\delta_{ij}$ as usual: but now $p_i \neq mv_i$.

This leads to the novel situation that the velocities in different directions *do not commute*. From

$mv_i = -i\hbar\nabla_i - qA_i / c$

it is easy to check that

$$[v_x, v_y] = \frac{iq\hbar}{m^2 c} B$$

To actually solve Schrödinger's equation for an electron confined to a plane in a uniform perpendicular magnetic field, it is convenient to use the Landau gauge,

$$\vec{A}(x,y,z)=(-By,0,0)$$

giving a constant field B in the z direction. The equation is

$$H\psi(x,y)=\left[\frac{1}{2m}(p_x+qBy/c)^2+\frac{p_y^2}{2m}\right]\psi(x,y)=E\psi(x,y).$$

Note that x *does not appear in this Hamiltonian,* so it is a cyclic coordinate, and p_x is conserved. In other words, this H commutes with p_x, so H and p_x have a common set of eigenstates. We know the eigenstates of p_x are just the plane waves $e^{ip_x x/\hbar}$, so the common eigenstates must have the form:

$$\psi(x,y)=e^{ip_x x/\hbar}\chi(y)$$

Operating on this wavefunction with the Hamiltonian, the operator p_x appearing in H simply gives its eigenvalue. That is, the p_x in H just becomes a number! Therefore, writing $p_y=-i\hbar d/dy$, the y-component $\chi(y)$ of the wavefunction satisfies:

$$-\frac{\hbar^2}{2m}\frac{d^2}{dy^2}\chi(y)+\tfrac{1}{2}m\left(\frac{qB}{mc}\right)^2(y-y_0)^2\chi(y)=E\chi(y)$$

where

$$y_0=-cp_x/qB$$

We now see that the conserved canonical momentum p_x in the x-direction is actually the coordinate of the center of a simple harmonic oscillator potential in the y-direction! This simple harmonic oscillator has frequency $\omega=|q|B/mc$, so the allowed values of energy for a particle in a plane in a perpendicular magnetic field are:

$$E=(n+\tfrac{1}{2})\hbar\omega=(n+\tfrac{1}{2})\hbar|q|B/mc.$$

The frequency is of course the cyclotron frequency that of the classical electron in a circular orbit in the field.

Let us confine our attention to states corresponding to the lowest oscillator state,. How many such states are there? Consider a square of conductor, area $A = L_x \times L_y$, and, for simplicity, take periodic boundary conditions. The center of the oscillator wave function y_0 must lie between 0 and L_y. But remember that $y_0 = -cp_x / qB$, and with periodic boundary conditions $e^{ip_x L_x/\hbar} = 1$, so $p_x = 2n\pi\hbar / L_x = nh / L_x$. This means that y_0 takes a series of evenly-spaced discrete values, separated by

$$\Delta y_0 = ch / qBL_x.$$

So the total number of states $N = L_y / \Delta y_0$,

$$N = \frac{L_x L_y}{\left(\frac{hc}{qB}\right)} = A \cdot \frac{B}{\Phi_0},$$

where ϕ_0 is called the "flux quantum". So the total number of states in the lowest energy level $E = \frac{1}{2}\hbar\omega$ (usually referred to as the lowest Landau level, or *LLL*) is exactly equal to the total number of flux quanta making up the field *B* penetrating the area *A*.

It is instructive to find y_0 from a purely classical analysis.

Writing $m\dot{\vec{v}} = \frac{q}{c}\vec{v} \times \vec{B}$ in components,

$$m\ddot{x} = \frac{qB}{c}\dot{y},$$

$$m\ddot{y} = -\frac{qB}{c}\dot{x}.$$

These equations integrate trivially to give:

$$m\dot{x} = \frac{qB}{c}(y - y_0),$$

$$m\dot{y} = -\frac{qB}{c}(x - x_0).$$

Here (x_0, y_0) are the coordinates of the center of the classical circular motion (the velocity vector $\dot{\vec{r}} = (\dot{x}, \dot{y})$ is always perpendicular to $(\vec{r} - \vec{r}_0)$), and $\vec{r}_0$ is given by

$$y_0 = y - cmv_x / qB = -cp_x / qB$$
$$x_0 = x + cmv_y / qB = x + cp_y / qB.$$

(Recall that we are using the gauge $\vec{A}(x, y, z) = (-By, 0, 0)$, and $p_x = \frac{\partial L}{\partial \dot{x}} = mv_x + \frac{q}{c} A_x$, etc.)

Chapter 3

Tensor Operators

REVIEW OF THE ROTATION OPERATOR

Recall that the matrix elements of the rotation operator

$$U\left(R\left(\vec{\theta}\right)\right)=e^{-\frac{i\vec{\theta}\cdot\vec{J}}{\hbar}}$$

within a definite j subspace are written

$$D^{(j)}_{m'm}\left(R\left(\vec{\theta}\right)\right)=\left\langle j,m'\right|e^{-\frac{i\vec{\theta}\cdot\vec{J}}{\hbar}}\left|j,m\right\rangle.$$

ROTATING A KET

By a ket, we mean here an element of the j subspace expressed as a sum over the basis kets $|j, m>$ (which we'll just write as $|m>$ here) with a set of coefficients a_m,

$$|\alpha\rangle=\sum_{m=-j}^{j}\alpha_m|m\rangle$$

and we are finding how that *set of coefficients* transforms under rotation. (Any such ket will stay within the j subspace under rotation, because the rotation operator commutes with $\vec{j}^2$.)

Rotating the ket by $\vec{\theta}$ to $|\alpha'\rangle=\sum\alpha'_{m'}|m'\rangle$,

$$|\alpha'\rangle=e^{-\frac{i\vec{\theta}\cdot\vec{J}}{\hbar}}|\alpha\rangle$$

$$=\sum_{m',m}\alpha_m|m'\rangle\langle m'|e^{-\frac{i\vec{\theta}\cdot\vec{J}}{\hbar}}|m\rangle$$

$$= \sum_{m',m} D^{(j)}_{m'm} \alpha_m |m'\rangle$$

So the ket rotation transformation is

$$\alpha'_{m'} = \sum_{m',m} D^{(j)}_{m'm} \alpha_m, \quad \text{or } \alpha' = D\alpha$$

with the usual matrix-multiplication rules.

Rotating a Basis Vector $|j,m>$

Now we look at something slightly different: suppose we apply the rotation operator to one of the *basis vectors* $|j,m>$, what is the result?

$$e^{-\frac{i\vec{\theta}\cdot\vec{J}}{\hbar}}|j,m\rangle = \sum_{m'}|j,m'\rangle\langle j,m'|e^{-\frac{i\vec{\theta}\cdot\vec{J}}{\hbar}}|j,m\rangle = \sum_{m'}|j,m'\rangle D^{(j)}_{m'm}(R).$$

Notice the reversal of *m, m′* compared with the operation on the set of component coefficients of the ket!

ROTATING AN OPERATOR

Just as in the Schrödinger versus Heisenberg formulations, we can either apply the rotation operator to the kets and leave the operators alone, or we can leave the kets alone, and rotate the operators:

$$A \rightarrow e^{\frac{i\vec{\theta}\cdot\vec{J}}{\hbar}} A e^{-\frac{i\vec{\theta}\cdot\vec{J}}{\hbar}} = U^{\dagger} A U$$

which will yield the same matrix elements, so the same physics.

Confusingly, there is a slightly different situation in which we need to rotate an operator, and it gives an opposite result. Suppose an operator T acts on a ket $|\alpha>$ to give the ket $T|\alpha>$. For kets $|\alpha>$ and $T|\alpha>$ to go to $U|\alpha>$ and $UT|\alpha>$ respectively under a rotation U, T itself must transform as $T \rightarrow UTU^{\uparrow}$ (recall $U^{\uparrow} = U^{-1}$).

SCALAR OPERATORS

A scalar operator is an operator which is invariant under rotations, for example the Hamiltonian of a particle in a spherically symmetric potential. (There are many less trivial examples of scalar operators, such as the dot product of two vector operators, as in a spin-orbit coupling.)

The transformation of an operator under an infinitesimal rotation is given by:

$$S \to U^{\dagger}(R)SU(R) \text{ with } U(R) = 1 - \frac{i\varepsilon \mathbf{J}\cdot\hat{\mathbf{n}}}{\hbar}$$

from which

$$S \to S + \left[\frac{i\varepsilon \mathbf{J}\cdot\hat{\mathbf{n}}}{\hbar}, S\right].$$

It follows that a scalar operator *S*, which does not change at all, must commute with all the components of the angular momentum operator, and hence must have a common set of eigenkets with, say, J^2 and J_z.

VECTOR OPERATORS

We have of course been using vector operators all along, for example $\vec{x}, \vec{L}, \vec{S}$ and so on. A vector operator has three components, corresponding to the Cartesian *x*, *y*, and *z* axes. Classically, such a three-dimensional vector transforms under rotation as

$$V_i \to \sum R_{ij} V_j,$$

with the usual rotation matrix, for example

$$R_z(\theta) = \begin{pmatrix} \cos\theta & -\sin\theta & 0 \\ \sin\theta & \cos\theta & 0 \\ 0 & 0 & 1 \end{pmatrix}$$

for rotation about the *z*-axis.

A quantum mechanical vector operator $\vec{V}$ is *defined* by requiring that the expectation values of its three components in any state transform like the components of a classical vector under rotation. It follows from this that the operator itself must transform vectorially,

$$V_i' = U^{\dagger}(R)V_iU(R) = \sum R_{ij}V_j$$

To see what this implies, it is easiest to look at a simple case. For an infinitesimal rotation about the *z*-axis,

$$R_z(\varepsilon) = \begin{pmatrix} 1 & -\varepsilon & 0 \\ \varepsilon & 1 & 0 \\ 0 & 0 & 1 \end{pmatrix}$$

the vector transforms

$$\begin{pmatrix} V_x \\ V_y \\ V_z \end{pmatrix} \to \begin{pmatrix} 1 & -\varepsilon & 0 \\ \varepsilon & 1 & 0 \\ 0 & 0 & 1 \end{pmatrix} \begin{pmatrix} V_x \\ V_y \\ V_z \end{pmatrix} = \begin{pmatrix} V_x - \varepsilon V_y \\ V_y + \varepsilon V_x \\ V_z \end{pmatrix}$$

The unitary Hilbert space operator U corresponding to this rotation $U(R_z(\varepsilon)) = 1 - \frac{i\varepsilon J_z}{\hbar}$, so

$$U^\dagger V_i U = (1 + i\varepsilon J_z / \hbar) V_i (1 - i\varepsilon J_z / \hbar)$$

$$= V_i + \frac{i\varepsilon}{\hbar}[J_z, V_i].$$

The requirement that the two transformations above, the infinitesimal classical rotation generated by $R_z(\varepsilon)$ and the infinitesimal unitary transformation $U^\uparrow(R) V_i U(R)$, are in fact the same thing yields the commutation relations of a vector operator with angular momentum:

$$i[J_z, V_x] = -\hbar V_y$$

$$i[J_z, V_y] = +\hbar V_x.$$

From this result and its cyclic equivalents, the components of *any* vector operator $\vec{v}$ must satisfy:

$$[V_i, J_j] = i\varepsilon_{ijk}\hbar V_k.$$

CARTESIAN TENSOR OPERATORS

The standard definition of a *Cartesian tensor* is that each of its suffixes transforms under rotation as do the components of an ordinary three-dimensional vector. Specifically, the nine components of a second rank tensor transform as:

$$T_{ij} \to T'_{ij} = \sum\sum R_{ii'} R_{jj'} T_{i'j'}.$$

A particular example of a second-rank tensor is $T_{ij} = U_i V_j$, where $\vec{U}$ and $\vec{V}$ are ordinary three-dimensional vectors.

The problem with this tensor is that it is *reducible* combinations of the elements can be arranged in groups such that rotations operate only within these groups. This is made evident by writing:

$$U_i V_j = \frac{\vec{U}\cdot\vec{V}}{3}\delta_{ij} + \frac{(U_i V_j - U_j V_i)}{2} + \left(\frac{U_i V_j + U_j V_i}{2} - \frac{\vec{U}\cdot\vec{V}}{3}\delta_{ij}\right).$$

The first term, the dot product of the two vectors, is clearly a *scalar* under rotation, the second term, which is an antisymmetric tensor has three independent components which are the *vector* components of the vector product $\vec{U}\times\vec{V}$, and the third term is a *symmetric traceless tensor*, which has five independent components. Altogether, then, there are 1 + 3 + 5 = 9 components, as required.

SPHERICAL TENSORS

Notice the numbers of elements of these irreducible subgroups: 1, 3, 5. These are exactly the numbers of elements of angular momenta representations for $j = 0, 1, 2$! The first term is of course trivial: the scalar by definition is not affected by rotation, and neither is an $j = 0$ state.

To deal with the second and third terms, we introduce tensor operators having three and five components, such that under rotation these sets of components transform among themselves just as do the sets of eigenkets of angular momentum in the $j = 1$ and $j = 2$ representations respectively.

Recall that the matrix elements of the rotation operator

$$U\left(R\left(\vec{\theta}\right)\right) = e^{-\frac{i\vec{\theta}\cdot\vec{J}}{\hbar}}$$

within a definite j subspace are written

$$D^j_{m'm}\left(R\left(\vec{\theta}\right)\right) = \langle j,m'|e^{-\frac{i\vec{\theta}\cdot\vec{J}}{\hbar}}|j,m\rangle$$

so under rotation operator a basis state $|j,m>$ transforms as:

$$e^{-\frac{i\vec{\theta}\cdot\vec{J}}{\hbar}}|j,m\rangle=\sum_{m'}|j,m'\rangle\langle j,m'|e^{-\frac{i\vec{\theta}\cdot\vec{J}}{\hbar}}|j,m\rangle=\sum_{m'}|j,m'\rangle D^{(j)}_{m'm}(R).$$

Notice the reversal of *m, m′* compared with the usual operation of a matrix on a vector: this is because we are rotating the basis here, rather than rotating relative to the basis.

The basic idea is that these irreducible subgroups into which Cartesian tensors apparently decompose under rotation (generalizing from our one example) form a more natural basis set of tensors for problems with rotational symmetries.

Definition: We define a spherical tensor of rank k as a set of $2k+1$ components T_k^q, $q=k,k-1,....,-k$ such that under rotation they transform like a set of angular momentum eigenkets, that is,

$$U(R)T_k^q U^\dagger(R)=\sum_{q'}D^{(k)}_{q'q}T_k^{q'}$$

To see the properties of these spherical tensors, it is useful to evaluate the above equation for infinitesimal rotations, for which, $D^{(k)}_{q,q}(\vec{\varepsilon})=\langle k,q'|l-i\vec{\varepsilon}.\vec{j}/\hbar|k,q\rangle$ to find

$$\left[J_\pm,T_k^q\right]=\pm\hbar\sqrt{(k\mp q)(k\pm q+1)}\,T_k^{q\pm1}$$
$$\left[J_z,T_k^q\right]=\hbar q T_k^q.$$

This set of commutation relations could be taken as the *definition* of the spherical tensors.

A SPHERICAL VECTOR

The $j=1$ angular momentum eigenkets are just the familiar spherical harmonics

$$Y_1^0=\sqrt{\frac{3}{4\pi}}\frac{z}{r},\quad Y_1^{\pm1}=\mp\sqrt{\frac{3}{4\pi}}\frac{x\pm iy}{\sqrt{2}r}.$$

The rotation operator will transform (x, y, z) as an ordinary vector in three-space, and this is evidently equivalent to

$$|j=1,m\rangle \to \sum_{m'} |j=1,m'\rangle D^{(j)}_{m'm}(R)$$

It follows that the spherical representation of a three vector (v_x, v_y, v_z) has the form:

$$T_1^{\pm 1} = \mp \frac{V_x \pm iV_y}{\sqrt{2}} = V_1^{\pm 1}, \quad T_1^0 = V_z = V_1^0.$$

In line with spherical tensor notation, the components (T_1^1, T_1^0, T_1^{-1}) are denoted T_1^a.

THE WIGNER-ECKART THEOREM

At this point, we must bear in mind that these tensor operators are not necessarily just functions of angle. For example, the position operator is a spherical vector multiplied by the radial variable *r*, and kets specifying atomic eigenstates will include radial quantum numbers as well as angular momentum.

Therefore, the matrix element of a tensor between two states will look like $\langle a_2, j_2, m_2 \mid T_k^q \mid a_1, j_1, m_1 \rangle$, where the *j*'s and *m*'s denote the usual angular momentum eigenstates and the *a*'s are nonangular quantum numbers, such as radial states.

The basic point of the Wigner–Eckart theorem is that the angular dependence of these matrix elements can be factored out, and that it is given by the Clebsch-Gordan coefficients. Having factored it out, the remaining dependence, which is only on the total angular momentum in each of the kets, not the relative orientation, and of course on the a's, is traditionally written as a bracket with double lines, that is,

$$\langle \alpha_2, j_2, m_2 | T_k^q | \alpha_1, j_1, m_1 \rangle = \langle \alpha_2, j_2 \| T_k \| \alpha_1, j_1 \rangle \cdot \langle j_2, m_2 | k, q; j_1, m_1 \rangle.$$

for example, and is not that difficult. The basic strategy is to put the defining identities.

$$\left[J_{\pm}, T_k^q\right] = \pm\hbar\sqrt{(k \mp q)(k \pm q + 1)}\, T_k^{q\pm 1}$$
$$\left[J_z, T_k^q\right] = \hbar q T_k^q$$

between $|a,k,m>$ bras and kets, then get rid of the $J_{\pm}$ and J_z by having them operate on the bra or ket. This generates a series of linear equations for $\left\langle \alpha_2, j_2, m_2 \,|\, T_k^q \,|\, \alpha_1, j_1, m_1 \right\rangle$ matrix elements with m variables differing by one, and in fact this set of linear equations is *identical* to the set that generates the Clebsch-Gordan coefficients, so we must conclude that these spherical tensor matrix elements, ranging over possible m and j values, are exactly proportional to the Clebsch-Gordan coefficients and that is the theorem.

A Few Hints for 15.3.3: that first matrix element comes from adding a spin j to a spin 1, writing the usual maximum m state, applying the lowering operator to both sides to get the total angular momentum $j + 1$, $m = j$ state, then finding the same m state orthogonal to that, which corresponds to total angular momentum j (instead of $j + 1$).

For the operator J, the Wigner-Eckart matrix element simplifies because J cannot affect a, and also it commutes with J^2, so cannot change the total angular momentum.

So, in the Wigner-Eckart equation, replace T_k^q on the left-hand side by J_1^0, which is just J_z. The result of (1) should follow.

(2) First note that a scalar operator cannot change m. Since c is independent of A we can take $A = J$ to find c.

Chapter 4

Variational Methods

VARIATIONAL METHOD FOR FINDING THE GROUND STATE ENERGY

The idea is to guess the ground state wave function, but the guess must have an adjustable parameter, which can then be varied (hence the name) to minimize the expectation value of the energy, and thereby find the best approximation to the true ground state wave function. This crude sounding approach can in fact give a surprisingly good approximation to the ground state energy (but usually not so good for the wave function, as will become clear).

We'll begin with a single particle in a potential, $H = p^2/2m + V(\vec{r})$. If the particle is restricted to one dimension, and we're looking for the ground state in any fairly localized potential well, we can start with the family of normalized Gaussians, $|\psi,\alpha> = \left(\frac{\alpha}{\pi}\right)^{1/4} e^{-\alpha x^2/2}$: just find $\langle\psi,\alpha|H|\psi,\alpha\rangle$, differentiate the result with respect to, setting this to zero (and checking that you have in fact found a minimum.) Not surprisingly, this gives the exact ground state for the simple harmonic oscillator potential, and for nothing else. What is perhaps surprising is that the result is only off by 30per cent or so for the attractive delta-function potential, even though the wave function looks a lot different.Obviously, the Gaussian family cannot be used if there is an infinite wall anywhere: one must find a family of wave functions vanishing at the wall.

To gain some insight into what we're doing, suppose the Hamiltonian $H = p^2/2m + v(\vec{r})$ has the set of (unknown to us) eigenstates

$$H|n>= E_n|n>.$$

Since the Hamiltonian is Hermitian, these states span the space of possible wave functions, including our variational family, so:

$$|\psi,\alpha>= \Sigma a_n(\alpha)|n>.$$

From this,

$$\frac{\langle\psi,\alpha|H|\psi,\alpha\rangle}{\langle\psi,\alpha|\psi,\alpha\rangle} = \sum |a_n|^2 E_n \geq E_0$$

for *any*$|\psi,\alpha>$. (We don't need the denominator if we've chosen a family of normalized wave functions, as we did with the Gaussians above.) Evidently, minimizing $\frac{\langle\psi,\alpha|H|\psi,\alpha\rangle}{\langle\psi,\alpha|\psi,\alpha\rangle}$ as a function of α gives us an upper bound on the ground state energy, hopefully not too far from the true value.

We can see immediately that this will probably be better for finding for the ground state energy than for mapping the ground state wave function: suppose the optimum state in our family is actually$|\alpha_{min}\rangle = N(|0\rangle + 0.2|1\rangle)$,with the normalization constant $N \cong 0.098$, a 20per cent admixture of the first excited state. Then the wave function is off by of order 20per cent, but the energy estimate will be too high by $0.04(E_1 - E_0)$ usually a much smaller error.

To get some idea of how well this works, Messiah applies the method to the ground state of the hydrogen atom. We know it's going to be spherically symmetric, so it amounts to a one-dimensional problem: just the radial wave function. Using standard notation,

$$a_0 = \hbar^2 / me^2,\ E_0 = me^4 / 2\hbar^2, \quad \rho = r / a_0$$

and for a trial wave function u

$$E(u) = -E_0 \frac{\int u\left(\frac{d^2}{d\rho^2} + \frac{2}{\rho}\right) u\, d\rho}{\int u^2 d\rho}$$

(we're going to take u real).

Messiah tries three families:

$$u_1 = \rho e^{-\alpha\rho}$$

$$u_2 = \frac{\rho}{\alpha^2 + \rho^2}$$

$$u_3 = \rho^2 e^{-\alpha\rho}$$

and finds $\alpha_{min} = 1, \pi/4, 3/2$ respectively. The first family, u_1, includes the exact result, and the minimization procedure finds it.

For the three families, then the energy of the best state is off by 0, 25per cent, 21per cent respectively.

The wave function error is defined as how far the square of the overlap with the true ground state wave function falls short of unity. For the three families, $\varepsilon = 1 - |\langle \psi_0 | \psi_{var} \rangle|^2$:0, 0.21, 0.05. Notice here that our hand waving argument that the energies would be found much more accurately than the wave functions comes unstuck. The third family has far better wave function overlap than the second, but only a slightly better energy estimate. Why? A key point is that the potential is singular at the origin, there is a big contribution to potential energy from a rather small region, and the third family wave function is the least accurate of the three there. The second family functions are very inaccurate at large distances: the expectation value $\langle r \rangle = 1.5\alpha_0, \infty, 1.66\alpha_0$ for the three families.

But at large distances, both kinetic and potential energies are small, so the result can still look reasonable. These examples reinforce the point that the variational method should be used cautiously.

VARIATIONAL METHOD FOR HIGHER STATES

In some cases, the approach can be used easily for higher states: specifically, in problems having some symmetry. For example, if the one dimensional attractive potential is symmetric about the origin, and has more than one bound state, the ground state will be even, the first excited state odd. Therefore, we can estimate the energy of the first excited state by minimizing a family of odd functions, such as

$$\psi(x,\alpha)=\left(\sqrt{\pi}/2\alpha^{3/2}\right)xe^{-\alpha x^2/2}$$

GROUND STATE ENERGY OF THE HELIUM ATOM BY THE VARIATIONAL METHOD

We know the ground state energy of the hydrogen atom is –1 Ryd, or –13.6 ev. The He^+ ion has Z = 2, so will have ground state energy, proportional to Z^2, equal to –4 Ryd. Therefore for the He atom, if we neglect the electron-electron interaction, the ground state energy will be -8 Ryd, –109 ev., the two electrons having opposite spins will both be in the lowest spatial state. Actually, experimentally, the He atom ground state energy is only –79 ev, because the repulsion between the electrons loosens things.

To get a better value for the ground state energy still using tractable wave functions, we change the wave functions from the ionic wave function $\left(Z^3/\pi a_0^3\right)^{1/2} e^{-Zr/a_0}$ with Z = 2 $\left(Z'^3/\pi a_0^3\right)^{1/2} e^{-Zr/a_0}$ to with Z' now a variable parameter. In other words, we are trying to allow for the electron-electron repulsion, which must push the wave functions out a bit, by keeping exactly the same shaped wave function but lessening the effective nuclear charge as reflected in the spread of the wave function from Z to Z', and we'll determine Z' by varying it to find the minimum total energy, including the term from electron-electron repulsion.

To find the potential energy from the nuclear-electron interactions, we of course use the actual nuclear charge $Z = 2$, but the Z' wave function, so the nuclear *P.E.* for the two electrons is:

$$P.E. = -2Ze^2 \int_0^\infty \frac{1}{r} 4\pi r^2 dr \left(Z'^3 / \pi a_0^3\right) e^{-2Z'r/a_0}$$

$$= -4ZZ' \left(e^2 / 2a_0\right)$$

$$= -8Z' \text{ Ryd} \quad (Z = 2).$$

This could have been figured out from the formula for the one-electron ion, where the potential energy for the one electron is $-2Z^2$ Ryd, one factor of Z being from the nuclear charge, the other from the consequent shrinking of the orbit.

The kinetic energy is even easier: it depends entirely on the shape of the wave function, not on the actual nuclear charge, so for our trial wave function it has to be Z'^2 Ryds per electron.

The tricky part is the *P.E.* for the electron-electron interaction. This is positive.

Each electron has a wave function $\left(Z'^3 / \pi a_0^3\right)^{1/2} e^{-Zr/a_0}$, a spherical charge probability distribution.

Denoting charge probability density by $\rho(r)$, we need

$$I = \iint d\vec{r}_1 d\vec{r}_2 \frac{\rho(\vec{r}_1)\rho(\vec{r}_2)}{|\vec{r}_1 - \vec{r}_2|}$$

$$= 16\pi^2 \int_0^\infty r_1^2 dr_1 \int_0^\infty r_2^2 dr_2 \frac{\rho(r_1)\rho(r_2)}{r_>},$$

$$r_> = \max(r_1, r_2).$$

This integral must be done in two stages:

$$I = 16\pi^2 \int_0^\infty \rho(r_1) r_1 dr_1 \left(\int_0^{r_1} \rho(r_2) r_2^2 dr_2 + r_1 \int_{r_1}^\infty \rho(r_2) r_2 dr_2 \right).$$

The result is -(5/4)Z′ Collecting terms, the total energy (for $Z = 2$) is: $E = -2\left(4Z' - Z'^2 - \frac{5}{8}Z'\right)$ Ryd

and this is minimized by taking $Z' = 2 - \frac{5}{16}$, giving an energy of –77.5 ev, off the true value by about 1 ev, so indeed the presence of the other electron is taken care of as far as total energy is concerned by shielding the nuclear charge by an amount (5/16)*e*.

Chapter 5

Time-Independent Perturbation Theory

INTRODUCTION

If an atom (not necessarily in its ground state) is placed in an external electric field, the energy levels shift, and the wave functions are distorted. This is called the *Stark effect*. The new energy levels and wave functions could in principle be found by writing down a complete Hamiltonian, including the external field, and finding the eigenkets. This actually can be done in one case: the hydrogen atom, but even there, if the external field is small compared with the electric field inside the atom (which is billions of volts per meter) it is easier to compute the changes in the energy levels and wave functions with a scheme of successive corrections to the zero-field values. This method, termed *perturbation theory,* is the single most important method of solving problems in quantum mechanics, and is widely used in atomic physics, condensed matter and particle physics.

It should be noted that there *are* problems which cannot be solved using perturbation theory, even when the perturbation is very weak, although such problems are the exception rather than the rule. One such case is the one-dimensional problem of free particles perturbed by a localized potential of strength λ. As we found earlier in the course, switching on an arbitrarily weak attractive potential causes the free particle wave function to drop below the continuum of plane wave energies and become a localized bound state with binding energy of order λ^2. However, changing the sign of λ to give a repulsive potential there is no bound state, the lowest energy plane wave state stays at energy zero. Therefore the energy shift on switching on the perturbation cannot be represented as a power series in λ, the strength of the perturbation. This particular difficulty does not in general occur in three dimensions, where arbitrarily weak potentials do not give bound states except for certain many-body problems (like the Cooper pair problem) where the exclusion principle reduces the effective dimensionality of the available states.

THE PERTURBATION SERIES

We begin with a Hamiltonian H^0 having known eigenkets and eigenenergies:

$$H^0 |n^0\rangle = E_n^0 |n^0\rangle.$$

The task is to find how these eigenkets and eigenenergies change if a small term H^1 (an external field, for example) is added to the Hamiltonian, so:

$$\left(H^0 + H^1\right)|n\rangle = E_n |n\rangle.$$

That is to say, on switching on H^1,

$|n^0\rangle \to |n\rangle,\ E_n^0 \to E_n.$

The basic assumption in perturbation theory is that H^1 is sufficiently small that the leading corrections are the same order of magnitude as H^1 itself, and the true energies can be better and better approximated by a successive series of corrections, each of order H^1/H^0 compared with the previous one.

The strategy, then, is to expand the true wave function and corresponding eigenenergy as series in H^1/H^0. These series are then fed into $\left(H^0 + H^1\right)|n> = E_n|n>$, and terms of the same order of magnitude in H^1/H^0 on the two sides are set equal. The equations thus generated are solved one by one to give progressively more accurate results.

To make it easier to identify terms of the same order in H^1/H^0 on the two sides of the equation, it is convenient to introduce a dimensionless parameter λ which always goes with H^1, and then expand $|n>, E_n$ as power series in $\lambda, |n\rangle = |n^0\rangle + \lambda|n^1\rangle + \lambda^2|n^2\rangle + \ldots$, etc. The ket $|n^m>$ multiplied by λ^m is therefore of order $\left(H^1/H^0\right)^m$.

This λ is purely a bookkeeping device: we will set it equal to 1 when we are through! It's just there to keep track of the orders of magnitudes of the various terms.

Putting the series expansions for $|n>, E_n$ in

$$\left(H^0 + \lambda H^1\right)|n\rangle = E_n|n\rangle$$

we have

$$\left(H^0 + \lambda H^1\right)\left(|n^0\rangle + \lambda|n^1\rangle + \lambda^2|n^2\rangle + \ldots\right) = \left(E_n^0 + \lambda E_n^1 + \lambda^2 E_n^2 + \ldots\right)\left(|n^0\rangle + \lambda|n^1\rangle + \lambda^2|n^2\rangle + \ldots\right).$$

We're now ready to match the two sides term by term in powers of λ.

The zeroth-order term, of course, just gives back $H^0|n^0\rangle = E_n^0|n^0\rangle$.

FIRST-ORDER TERMS

Matching the terms linear in *l* on both sides:

$$H^0|n^1\rangle + H^1|n^0\rangle = E_n^0|n^1\rangle + E_n^1|n^0\rangle.$$

This equation is the key to finding the first-order change in energy E_n^1. Taking the inner product of both sides with $\langle n^0|$:

$$\langle n^0|H^0|n^1\rangle + \langle n^0|H^1|n^0\rangle = \langle n^0|E_n^0|n^1\rangle + \langle n^0|E_n^1|n^0\rangle,$$

then using $\langle n^0 | H^0 = \langle n^0 | E_n^0$, and $\langle n^0 | n^0 \rangle = 1$, we find

$$E_n^1 = \langle n^0 | H^1 | n^0 \rangle.$$

Taking now $\lambda = 1$, we have established that the first-order change in the energy of a state resulting from adding a perturbing term to the Hamiltonian is just the expectation value of H^1 in that state.

For example, we can estimate the ground state energy of the helium atom by treating the electrostatic repulsion between the electrons as a perturbation. The zeroth-order ground state has the two (opposite spin) electrons in the ground state hydrogen-atom wave function (scaled for the doubling of nuclear charge). The first-order energy correction E_0^1 is then given by computing the expectation value $\langle e^2 / r_{12} \rangle$ for this ground state wave function.

The general expression for the first-order change in the *wave function* is found by taking the inner product of the first-order equation with the bra $\langle m^0 |, m \neq n$,

$$\langle m^0 | H^0 | n^1 \rangle + \langle m^0 | H^1 | n^0 \rangle = \langle m^0 | E_n^0 | n^1 \rangle + \langle m^0 | E_n^1 | n^0 \rangle.$$

The last term is zero, since $\langle m_0 | n_0 \rangle = 0$, and in the first term $\langle m^0 | H^0 = \langle m^0 | E_m^0$, so

$$\langle m^0 | n^1 \rangle = \frac{\langle m^0 | H^1 | n^0 \rangle}{E_n^0 - E_m^0}$$

and therefore the wave function correct to first order is:

$$|n\rangle = |n^0\rangle + |n^1\rangle = |n^0\rangle + \sum_{m \neq n} \frac{|m^0\rangle \langle m^0 | H^1 | n^0 \rangle}{E_n^0 - E_m^0}$$

THE SECOND-ORDER ENERGY TERM

To find the *second*-order correction to the energy, it is necessary to match the second-order terms in

$$\left(H^0+\lambda H^1\right)\left(\left|n^0\right\rangle+\lambda\left|n^1\right\rangle+\lambda^2\left|n^2\right\rangle+\ldots\right)=$$

$$\left(E_n^0+\lambda E_n^1+\lambda^2 E_n^2+\ldots\right)\left(\left|n^0\right\rangle+\lambda\left|n^1\right\rangle+\lambda^2\left|n^2\right\rangle+\ldots\right)$$

giving:

$$H^0\left|n^2\right\rangle+H^1\left|n^1\right\rangle=E_n^0\left|n^2\right\rangle+E_n^1\left|n^1\right\rangle+E_n^2\left|n^0\right\rangle.$$

Taking the inner product with $\left\langle n^0\right|$ yields:

$$\left\langle n^0\right|H^0\left|n^2\right\rangle+\left\langle n^0\right|H^1\left|n^1\right\rangle=E_n^0\left\langle n^0\middle|n^2\right\rangle+E_n^1\left\langle n^0\middle|n^1\right\rangle+E_n^2\left\langle n^0\middle|n^0\right\rangle.$$

The leading terms on the two sides cancel as before. What about the term $E_n^1\left\langle n^0 \mid n^1\right\rangle$? Since $|n\rangle=\left|n^0\right\rangle+\left|n^1\right\rangle$, and both $|n\rangle$, $\left|n^0\right\rangle$ are normalized, $\left\langle n^0\middle|n^1\right\rangle+\left\langle n^1\middle|n^0\right\rangle=0$ in leading order that is to say, $\left\langle n^0\middle|n^1\right\rangle$ is pure imaginary. That just means that if to this order $|n>$ has a component parallel to $|n^0>$, that component has a small pure imaginary amplitude, and $|n>$ can be written (to this order) as $|n\rangle=e^{i\alpha}\left|n^0\right\rangle+$ kets $\perp$ $\left|n^0\right\rangle$, with α small. But the phase factor can be eliminated by redefining the phase of $|n>$, and with that redefinition $|n1>$ has *no* component in the $|n^0>$ direction, we can therefore drop the term $E_n^1\left\langle n^0 \mid n^1\right\rangle$.

So the second-order correction to the energy is:

$$E_n^2=\left\langle n^0\right|H^1\left|n^1\right\rangle=\left\langle n^0\right|H^1\sum_{m\neq n}\frac{\left|m^0\right\rangle\left\langle m^0\right|H^1\left|n^0\right\rangle}{E_n^0-E_m^0}=\sum_{m\neq n}\frac{\left|\left\langle m^0\right|H^1\left|n^0\right\rangle\right|^2}{E_n^0-E_m^0}.$$

SELECTION RULES

Perturbation theory involves evaluating matrix elements of operators. Very often, many of the matrix elements in a sum are zero obvious tests are parity and the Wigner-Eckart theorem. These are examples of *selection rules*: tests to find if a matrix element may be nonzero.

THE QUADRATIC STARK EFFECT

When a hydrogen atom in its ground state is placed in an electric field, the electron cloud and the proton are pulled different ways, an electric dipole forms, and the overall energy is lowered.

The perturbing Hamiltonian from the electric field is $H' = e\varepsilon z = e\varepsilon r\cos\theta$, where ε is the electric field strength, the field is in the z-direction, the electron charge e is negative. We shall denote the unperturbed eigenenergies of the hydrogen atom by $E_n = E_{nlm} = -1/n^2$, so in particular we denote the ground state energy by E_1.

The first-order correction to the ground state energy $E_1^1 = \langle 100 | e\varepsilon z | 100\rangle$, where

$$|100\rangle \equiv \psi_{100}(r) = \left(\frac{1}{\pi a_0^3}\right)^{1/2} e^{-r/a_0}.$$

This first-order term is zero since there are equal contributions from positive and negative z.

The second-order term is

$$E_1^2 = \sum_{n\neq 1,l,m} \frac{|\langle nlm | e\mathcal{E}z | 100\rangle|^2}{E_1 - E_n}$$

where we are now using $| nlm >$ to denote the unperturbed hydrogen atom wave functions, and here the $E_n = -1/n^2$ (in Rydbergs) are the unperturbed energies.

Most of the terms in this infinite series are zero the selection rules help get rid of them as follows: since $e\varepsilon z$ is the m = 0 component of a spherical vector and $|100>$ is a zero angular momentum state, it follows from the Wigner-Eckart theorem that. $\langle nlm |$ can only be $\langle n10 |$ This reduces the second-order sum over states to:

$$E_1^2 = \sum_{n\neq 1} \frac{|\langle n10 | e\mathcal{E}z | 100\rangle|^2}{E_1 - E_n}.$$

This is still not easy to evaluate, but an *upper bound* can be found be observing that $|E_1 - E_n| \geq |E_1 - E_2|$, so

$$|E_0^2| < \frac{1}{E_2 - E_1}\sum_{n\neq 1}|\langle n10|e\mathcal{E}z|100\rangle|^2 = \frac{1}{E_2 - E_1}\sum_{n\neq 1,l,m}\langle 100|e\mathcal{E}z|nlm\rangle\langle nlm|e\mathcal{E}z|100\rangle$$

where we have temporarily *restored* the full sum over n, l, m, that is, we've put back all the zero terms. The reason for this seeming backward step is that, having taken the energy-difference denominator outside the sum, we can even include $|100\rangle$ in the $|n/m\rangle$ sum (it's another zero term) and in fact we can even include the plane-wave (ionized) states as well as the bound states, since the plane waves all have energy greater than zero. At this point, the $\sum_{n,l,m}$ sum becomes a sum over all states, and therefore just becomes the unit operator,

$$\sum_{n,l,m}|nlm\rangle\langle nlm| = I,$$

so $|E_1^2| < \frac{1}{E_2 - E_1}\langle 100|(e\mathcal{E}z)^2|100\rangle.$

For the ground state hydrogen wave function, $\langle 100|z^2|100\rangle = a_0^2$, $E_1 = -e^2/2a_0$, $E_2 = E_1/4$ so

$$|E_1^2| < \frac{1}{\left(\frac{3}{4}e^2/2a_0\right)}(e\mathcal{E})^2 a_0^2 = \frac{8}{3}\mathcal{E}^2 a_0^3.$$

Furthermore, since all the terms in the series for E_1^2 are negative, *the first term sets a lower bound on* $|E_1^2|$:

$$|E_1^2| > \frac{|\langle 210|e\mathcal{E}z|100\rangle|^2}{E_1 - E_2}.$$

This can be evaluated in straightforward fashion to find $|E_1^2| > 0.55 \times \frac{8}{3}\mathcal{E}^2 a_0^3$.

So, even though we have not actually evaluated the second-order correction to the energy explicitly, we have it bracketed between two values, the lower one being more than half the upper one. Other ingenious methods have been developed to find that the true answer is $|E_1^2| = \frac{9}{4}\mathcal{E}^2 a_0^3$, but in fact the whole problem can be solved exactly using parabolic coordinates.

DEGENERATE PERTURBATION THEORY: DISTORTED 2-D HARMONIC OSCILLATOR

The above analysis works fine as long as the successive terms in the perturbation theory form a convergent series. A necessary condition is that the matrix elements of the perturbing Hamiltonian must be smaller than the corresponding energy level differences of the original Hamiltonian. If H^0 has different states with the same energy, in other words degenerate energy levels, and the perturbation has nonzero matrix elements *between these degenerate levels,* then obviously the theory breaks down. To see just how it breaks down, and how to fix it, we consider the two-dimensional simple harmonic oscillator:

$$H^0 = \frac{p_x^2 + p_y^2}{2m} + \frac{1}{2} m\omega^2 \left(x^2 + y^2\right).$$

Recall that for the *one*-dimensional simple harmonic oscillator the ground state wave function is

$$|0\rangle = \left(\frac{m\omega}{\pi\hbar}\right)^{1/4} e^{-m\omega x^2/2\hbar} = \left(\frac{m\omega}{\pi\hbar}\right)^{1/4} e^{-\xi^2/2} \text{ with } \xi = \sqrt{\frac{m\omega}{\hbar}}x, \quad \text{and}$$

$$|1\rangle = \left(\frac{m\omega}{\pi\hbar}\right)^{1/4} \sqrt{2}\xi e^{-\xi^2/2}$$

The two-dimensional oscillator is simply a product of two one-dimensional oscillators, so, writing $\eta = \sqrt{\frac{m\omega}{\hbar}}y$, the ground state is $|0\rangle = \left(\frac{m\omega}{\pi\hbar}\right)^{1/2} e^{-(\xi^2+\eta^2)/2}$, and the two (degenerate) next states, energy above the ground state, are

$$|1,0\rangle = \left(\frac{m\omega}{\pi\hbar}\right)^{1/2} \sqrt{2}\xi e^{-(\xi^2+\eta^2)/2}, \quad |0,1\rangle = \left(\frac{m\omega}{\pi\hbar}\right)^{1/2} \sqrt{2}\eta e^{-(\xi^2+\eta^2)/2}.$$

Suppose now we add a small perturbation $H^1 = \alpha m\omega^2 xy$ with α a small parameter. Notice that $\langle 0|H^1|0\rangle = \langle 1,0|H^1|1,0\rangle = \langle 0,1|H^1|0,1\rangle = 0$, so according to naïve perturbation theory, there is *no* first-order correction to the energies of these states. However, on going to second-order in the energy correction, the theory breaks down. The matrix element $\left\langle 1,0\middle|H^1\middle|0,1\right\rangle$ is nonzero, but the two states $|0,1\rangle, |1,0\rangle$ have the same energy! This gives an infinite term in the series for E_n^2.

Yet we know that a small term of this type will not wreck a two-dimensional simple harmonic oscillator, so what is wrong with our approach? It is helpful to plot the original harmonic oscillator potential $\frac{1}{2}m\omega^2\left(x^2+y^2\right)$ together with the perturbing potential $\alpha m\omega^2 xy$. The first of course has circular symmetry, the second has axes in the directions $x = \pm y$, climbing most steeply from the origin along $x = y$, falling most rapidly in the directions $x = -y$. If we combine the two potentials into a single quadratic form, the original circles of constant potential become ellipses, with their axes aligned along $x = \pm y$.

The problem arises even in the classical two-dimensional oscillator: picture a ball rolling backwards and forwards in a smooth saucer, a circular bowl. Now imagine the saucer is made slightly elliptical. The ball will still roll backwards and forwards through the center if it is released along one of the axes of the ellipse, although with different periods, as the axes differ in steepness. However, if it is released at a point *off* the axes, it will describe a complex path resolvable into components in the two axis directions having different periods.

For the quantum oscillator as for the classical one, as soon as the perturbation is introduced, the eigenkets are in the direction of the new elliptic axes. This is a large change from the original x and y axes, and definitely *not* proportional to the small parameter. But the original unperturbed problem had circular symmetry, and there was no particular reason to choose the x and y axes as we did. If we had instead chosen as our original axes the lines $x = \pm y$, the kets would *not* have undergone large changes on switching on the perturbation.

The resolution of the problem is now clear: *before* switching on the perturbation, *choose a set of basis kets in a degenerate subspace such that the perturbation is diagonal in that subspace.*

In fact, for the simple harmonic oscillator example above, the problem can be solved exactly:

$$\frac{1}{2}m\omega^2\left(x^2+y^2\right)+\alpha m\omega^2 xy=\frac{1}{2}m\omega^2\left[(1+\alpha)\left(\frac{x+y}{\sqrt{2}}\right)^2+(1-\alpha)\left(\frac{x-y}{\sqrt{2}}\right)^2\right]$$

and it is clear that, despite the results of naïve first-order theory, there is indeed *a first order shift* in the energy levels,

$$\hbar\omega \to \hbar\omega\sqrt{1\pm\alpha} \approx \hbar\omega(1\pm\alpha/2)$$

THE LINEAR STARK EFFECT

The hydrogen atom, like the two-dimensional harmonic oscillator discussed above, has a nondegenerate ground state but degeneracy in its lowest excited states. Specifically, there are four $n = 2$ states, all having energy -1/4 Ryd:

$$\psi_{200}(r)=\left(\frac{1}{32\pi a_0^3}\right)^{1/2}\left(2-\frac{r}{a_0}\right)e^{-r/2a_0},$$

$$\psi_{210}(r,\theta,\phi)=\left(\frac{1}{32\pi a_0^3}\right)^{1/2}\left(\frac{r}{a_0}\right)e^{-r/2a_0}\cos\theta,$$

$$\psi_{21\pm1}(r,\theta,\phi)=\left(\frac{1}{32\pi a_0^3}\right)^{1/2}\left(\frac{r}{a_0}\right)e^{-r/2a_0}\sin\theta e^{\pm i\phi}$$

Perturbing this system with an electric field in the z-direction, $H' = e\varepsilon z = e\varepsilon r\cos\theta$, note first that naïve perturbation theory predicts *no* first-order shift in any of these energy levels. However, to second order, there is a nonzero matrix element between two degenerate levels $\langle 200|H'|210\rangle$.

All the other matrix elements between these basis kets in the four-dimensional degenerate subspace are zero, so the only diagonalization necessary is within the *two*-dimensional degenerate subspace spanned by $|200\rangle, |210\rangle$, where

$$H^1 = \begin{pmatrix} 0 & \Delta \\ \Delta & 0 \end{pmatrix}$$

with

$$\begin{aligned}\Delta &= \langle 200|H^1|210\rangle \\ &= e\mathcal{E}\left(\frac{1}{32\pi a_0^3}\right)\int_0^\infty\left(2-\frac{r}{a_0}\right)\left(\frac{r\cos\theta}{a_0}\right)^2 e^{-r/a_0}r^2 dr \sin\theta d\theta d\phi \\ &= -3e\mathcal{E}a_0.\end{aligned}$$

Diagonalizing H^1 within this subspace, then, the new basis states are $(|200\rangle \pm |210\rangle)/\sqrt{2}$ with energy shifts $\pm\Delta$, linear in the perturbing electric field.

The states $|21\pm1\rangle$ are not changed by the presence of the field to this approximation, so the complete energy map of the $n = 2$ states in the electric field has two states at the original energy of -1/4Ryd, one state moved up from that energy by Δ, and one down by Δ.

Notice that the new eigenstates $(|200\rangle \pm |210\rangle)/\sqrt{2}$ are *not* eigenstates of the parity operator a sketch of their wave functions reveals that in fact they have nonvanishing electric dipole moment $\vec{\mu}$, indeed this is the reason for the energy shift, $\pm\Delta = \mp 3e\varepsilon a_0 = \mp\vec{\mu}.\vec{\varepsilon}$.

Chapter 6

Electrons in One Dimension

INTRODUCTION: LOOKING FOR SUPERCONDUCTORS AND FINDING INSULATORS

In 1964, Little suggested that it might be possible to synthesize a room temperature superconductor using organic materials in which the electrons traveled along certain kinds of chains, effectively confined to one dimension.

The first satisfactory theory of "ordinary" superconductivity, that of Bardeen, Cooper and Schrieffer (BCS) had appeared a few years earlier, in 1957. The key point was that electrons became bound together in opposite spin pairs, and at sufficiently low temperatures these bound pairs, being boson like, formed a coherent condensate all the pairs had the same total momentum, so all traveled together, a supercurrent. The locking of the electrons into this condensate effectively eliminated the usual single-electron scattering by impurities that degrades ordinary currents in conductors.

But what could bind the electrostatically repelling electrons? The answer turned out to be lattice distortions, as first suggested by Fröhlich in 1950. An electron traveling through the crystal attracts the positive ions, the consequent excess of local positive charge attracts another electron. The strength of this binding, and hence the temperature at which the superconducting transition takes place, depends on the rapidity of the lattice response. This was confirmed by the isotope effect: lattice response time obviously depends on the inertia of the lattice, the BCS theory predicted that for a superconducting element with different isotopic varieties, the ratio of the superconducting transition temperatures for pure isotopes T_2/T_1 was equal to $\sqrt{M_1/M_2}$, M_1, M_2 being the ion masses, the lighter isotope having the higher transition temperature. This was indeed the case.

Little's idea was that the build up of positive charge by a passing electron could be speeded up dramatically if instead of having to move ions, it need only rearrange other electrons. Unfortunately, there were no obvious three-dimensional candidate materials. However, if the conduction electrons moved along a one-dimensional chain, polarizable side chains might be attached, and rearrangement of the electronic charge distribution in these side chains would respond very rapidly to a passing conduction electron, building up a local positive charge. If this worked, order of magnitude arguments suggested possible enhancement of the transition temperature by a factor $\sqrt{M/m}$ over ordinary superconductors, m being the electron mass.

In the 1970's, various organic materials were synthesized and tested, beginning with one called TTF-TCNQ, in which a set of polymer-like long molecules donated electrons to another set, leaving one-dimensional conductors with partially filled bands, seemingly good candidates for superconductivity. Unfortunately, on cooling these materials surprisingly became *insulators* rather than superconductors! This was the first example of a *Peierls transition*, a widespread phenomenon in quasi one-dimensional systems.

The basic mechanism of the Peierls transition can be understood with a simple model. It is a nice example of applied second-order perturbation theory, including the degenerate case. We examine the model and the result below.

It should be added that in some newer materials the Peierls transition is (unexpectedly) suppressed under high pressure, and superconductivity has in fact been observed in organic salts, but so far only at transition temperatures around one Kelvin: Little's dream is not yet realised.

SECOND-ORDER PERTURBATION THEORY

To understand how a one-dimensional conductor might turn into an insulator at low temperatures, we must first become familiar with the simplest model of a one-dimensional conductor: $H = H^0 + V = \frac{p^2}{2m} + V(x)$

with H^0 a gas of noninteracting electrons on a line, and V periodic, that is $V(x+a) = v(x)$,

the potential from a line of ions spaced *a* apart. We'll take the system to have *N* ions in a total length *L*, so

$L = Na$ and to keep the math simple, we'll require periodic boundary conditions.

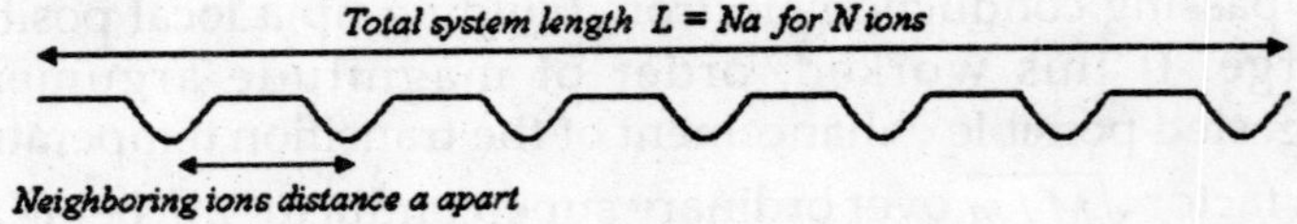

Ionic potential seen by electrons in one-dimensional system.

The physics here is that *without* the potential, the electron eigenstates are *plane waves*. The effect of the lattice potential is to partially reflect the waves, like a diffraction grating, generating components at different wavelengths. This effect becomes particularly important when the electron wavelength matches twice the ion spacing. For that case, the reflected and original waves have the same strength, the electron is at a standstill.

The eigenstates of H^0 are then

$|k\rangle^{(0)} = \frac{1}{\sqrt{L}} e^{ikx}$, with $e^{ikL} = 1$, so $k = \frac{2\pi n}{L}$, n being an integer.

The unperturbed energy eigenvalues, $H^0 \mid k_n\rangle^{(0)} = E_n^0 \mid k_n\rangle^{(0)}$,

are just $E^0 = \frac{\hbar^2 k^2}{2m}$

(*Note*: this is to be understood as.

$$H^0 \mid k\rangle^{(0)} = E_n^0 \mid k_n\rangle^{(0)}, \text{ with } E_n^0 = \frac{\hbar^2 k^2}{2m} \text{ and } k_n = \frac{2\pi n}{L}$$

We are following standard practice here. We shall also write. $\sum_k f(k)$ meaning $\sum_n f(k_n)$)

It's worth plotting the (E, k) curve:

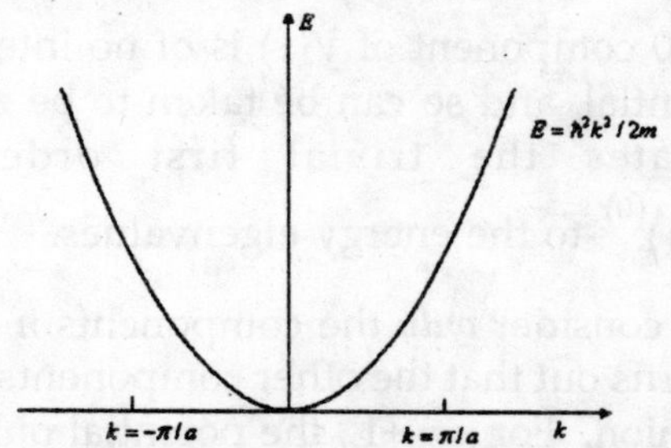

Energy momentum (E, k) curve for a free electron in one dimension.

Suppose we have ions with two electrons each to contribute to this one-dimensional (supposed) conductor. Assuming they move into these plane wave states, in the system ground state they will fill up the lowest energy states up to a maximum k-value denoted by $\pm k_F$ (F stands for Fermi, this is the *Fermi* momentum.) Where is it?

We know there will be a total of $2N$ electrons. We also know that the allowed values of k, from the boundary conditions, are $k_n = 2\pi n / L$, with n an integer. In other words, the allowed k's are uniformly spaced $2\pi n / L$ apart, meaning they have a *density* of $L / 2\pi$ in k-space, so the total number between $\pm k_F$ is Lk_F / π. The $2N$ electrons will have N of each spin, each k-state can take two electrons (one of each spin), so $Lk_F / \pi = N = L / a$, and $k_F = \pi / a$.

To do perturbation theory, we must find the matrix elements of $V(x)$ between eigenstates of H^0:

$$^{(0)}\langle k' | v | k \rangle^{(0)} = \frac{1}{L} \int e^{i(k-k')} V(x)\, dx$$

This is just the Fourier component v_{k-k} of $V(x)$.If $V(x)$ is periodic with period a,

$V_k \neq 0$ only if $k = nk$, n an integer, k = 2π/a In other words, if a function is periodic with spatial period *a*, the only nonzero Fourier components are those having the same spatial period *a*.

Therefore

$V(x) = \sum_n V_{nK} e^{inKx}$, and $V_{-nK} = V_{nK}^*$ since $V(x)$ is real; $K = 2\pi / a$.

The $n = 0$ component of $V(x)$ is of no interest it is just a constant potential, and so can be taken to be zero. Note that this eliminates the trivial first order correction $E_k^1 = ^{(0)}\langle k|V|k\rangle^{(0)}$ to the energy eigenvalues.

We shall consider *only* the components $n = +1$ and $n = -1$ of $V(x)$, it turns out that the other components can be treated in similar fashion. For $n = +1$,, the potential only has nonzero matrix elements between the plane wave state k and $k + k$, $k - k$ respectively.

So, the second order correction to the energy is:

$$E_k^2 = \sum_{k \neq k} \frac{\left|^{(0)}\langle k|V|k'\rangle^{(0)}\right|^2}{E_k^0 - E_{k'}^0}$$

$$= \frac{\left|^{(0)}\langle k|V|k+K\rangle^{(0)}\right|^2}{E_k^0 - E_{k+K}^0} + \frac{\left|^{(0)}\langle k|V|k-K\rangle^{(0)}\right|^2}{E_k^0 - E_{k-K}^0}$$

$$= \frac{|V_K|^2}{E_k^0 - E_{k+K}^0} + \frac{|V_{-K}|^2}{E_k^0 - E_{k-K}^0}.$$

This result is reasonable provided the terms are small, that is, the energy differences appearing in the denominators are large compared to the relevant Fourier component V_K. However, this cannot always be true! Notice that the state k = π/a has exactly the same unperturbed energy E^0 as the state $k - K = -\pi / a$: in this case, nondegenerate perturbation theory is clearly wrong. In fact, even for states close to, the energy denominator $E_k^0 - E_{k-k}^0$ is small compared with the numerator $|V_{-k}|^2$, so the series is not converging.

QUASI-DEGENERATE PERTURBATION THEORY NEAR THE CRITICAL WAVELENGTH

The good news is that, despite the many states near $k = \pi / a$ and $k = -\pi / a$ that are close together in energy, for any *one* state k near π / a the potential only has a nonzero matrix element to *one* other state close in energy, the state k – K, that is $k = -2\pi / a$. The strategy now is to do what might be called *quasidegenerate* perturbation theory: to diagonalize the full Hamiltonian in the subspace spanned by these two states $|k\rangle^{(0)}, |k-k\rangle^{(0)}$. Other states with nonzero matrix elements to these states are relatively much further away in energy, and can be treated using ordinary perturbation theory.

The matrix elements of the full Hamiltonian in the subspace spanned by these two states are:

$$\begin{vmatrix} E_k^0 & V_K^* \\ V_K & E_{k-K}^0 \end{vmatrix}.$$

Diagonalizing *within this subspace* gives energy eigenvalues:

$$E_\pm = \frac{1}{2}(E_k^0 + E_{k-K}^0) \pm \sqrt{\left(\frac{E_k^0 - E_{k-K}^0}{2}\right)^2 + |V_K|^2}.$$

Notice that, provided $\left|E_k^0 - E_{k-k}^0\right| >> |V_k|$, to leading order this gives back $E_\pm = E_k^0, E_{k-k}^0$ the order depending on k. However, as k approaches π / a, $\left|E_k^0 - E_{k-k}^0\right|$ becomes of order, and the energies deviate from the unperturbed values. If k is approaching from below, $E_k = E_- < E_k^0$, and the lower energy is pushed *downwards* by the perturbation: This is a common occurrence with almost degenerate states, perturbations cause the energy levels to "repel" each other.

For $\text{k} = \pi / a, E_{k-k}^0 = E_{k-2\pi/a}^0 = E_k^0$ At this value of k, the unperturbed states are exactly degenerate, and the perturbation lifts the degeneracy to give $E_\pm = E_{\pi/a}^0 \pm |V_k|$.

In the graph below, the green (continuous) curve is the unperturbed energy as a function of *k*, the red curve (with the step) the calculated energy including the leading correction from the periodic potential.

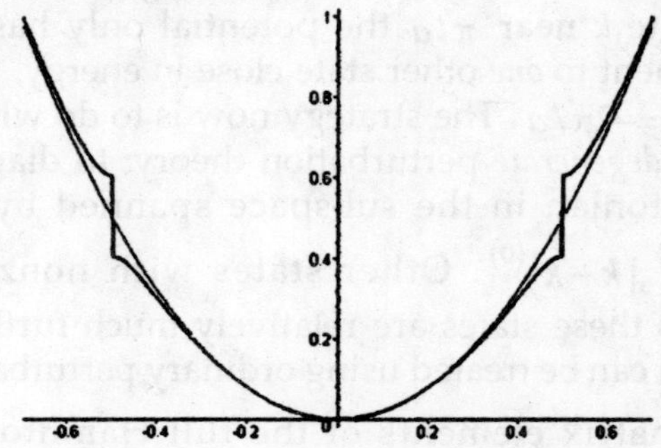

ENERGY GAPS AND BANDS

The energy jump, or gap, of $2|V_K|$ at $|k| = p/a$ means that there are no plane wave type eigenstates with energies in that range attempting to integrate Schrödinger's equation in the periodic potential for such an energy gives exponentially growing and decaying solutions. Such energy gaps in fact are present in real crystalline solids, the allowed energies are said to be in "bands". The lowest band for our model is from $k = -\pi/a$ to π/a Since the allowed values of *k* are given by $k = 2\pi n/L$, the spacing between adjacent *k*'s is $2\pi/L$ and the total number of *k*'s in the lowest band is $L/a = N$, the same as the number of atoms. Since each electron has two spin states, this implies that a one-dimensional crystal of *divalent* atoms will *just fill* the lowest band with electrons. Therefore, any outside field can only excite an electron to a different state if an energy of at least $2|V_K|$ is supplied for a small electric field, the filled band of electrons will remain in the ground state, there will be no current. This material is an insulator.

On the other hand, if *monovalent* atoms are used, it is clear that the lowest band is *only half full*, adjacent empty electron states are available. The electrons are free to accelerate if an external field is applied. Barring the unexpected, this one-dimensional crystal would be a metal.

Let us now examine how the periodic potential alters the eigenstates. Ignoring the small corrections from plane waves outside the $|k\rangle^{(0)}, |k-K\rangle^{(0)}$ subspace, the eigenstates to this order have the form

$$|k\rangle = a_k |k\rangle^{(0)} + a_{k-K} |k-K\rangle^{(0)}$$

where

$$\frac{a_{k-K}}{a_k} = \frac{E_- - E_k^0}{V_K^*}$$

from the diagonalization of the 2 × 2 matrix representing the Hamiltonian in the subspace.

As k increases from 0 towards, the plane wave initially proportional to e^{ikx} has a gradually increasing admixture of $e^{i(k-2\pi/a)x}$, until at $k = \pi/a$ the two have equal weight meaning that the eigenfunction is now a standing wave. In fact, there are *two* standing wave solutions at $k = \pi/a$, corresponding to the energies below and above the gap. Taking the atoms to have an attractive potential, the lower energy wave has a probability distribution peaking at the atomic positions. The diffractive scattering that gives a left-moving component to a right moving wave is known as Bragg scattering. It also manifests itself in the *group velocity* of the electronic excitations, $v_{group} = d\omega/dk = (1/\hbar)dE/dk$ An electron injected into a one-dimensional metal would not be a plane wave state, but a wavepacket traveling at the group velocity. It is evident that for an injected electron with mean value of k close to π/a, the electron will move very slowly into the metal. This is to be expected the eigenstates become standing waves as $k \rightarrow \pi/a$.

For three-dimensional crystals, the situation is far more complicated, but many of the same ideas are relevant. Electron waves are now diffracted by whole planes of atoms, and the three-dimensional momentum space is divided into Brillouin zones, with planes having an energy gap across them.

THE PEIERLS TRANSITION: HOW COOLING A CONDUCTOR CAN GIVE AN INSULATOR

As mentioned in the Introduction, substances very close to monovalent one-dimensional crystals have been synthesized, and it has been found surprisingly that at low temperatures many of them undergo a transition from metallic to *insulating* behaviour. *What happens is that the atoms in the lattice rearrange slightly, moving from an equally-spaced crystal to one in which the spacing alternates, that is, the atoms form pairs.* This is called *dimerization,* and costs some elastic energy, since for identical atoms the lowest state must be one of equal spacing for any reasonable potential. However, the *electrons* are able to move to a lower energy state by this maneuver.

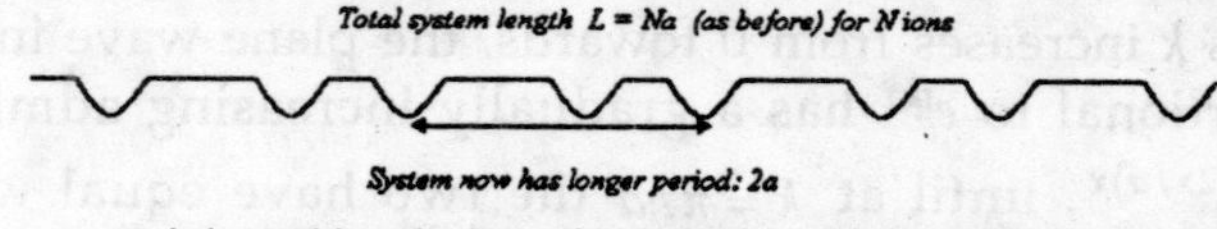

Ionic potential seen by electrons in a dimerized one-dimensional system.

Just how this happens can be understood using the perturbation theory analysis above. For equally spaced atoms, the electrons *half-fill* the band, that is, they fill it up (two electrons, one of each spin, per state) to $|k| = \pi / 2a$.

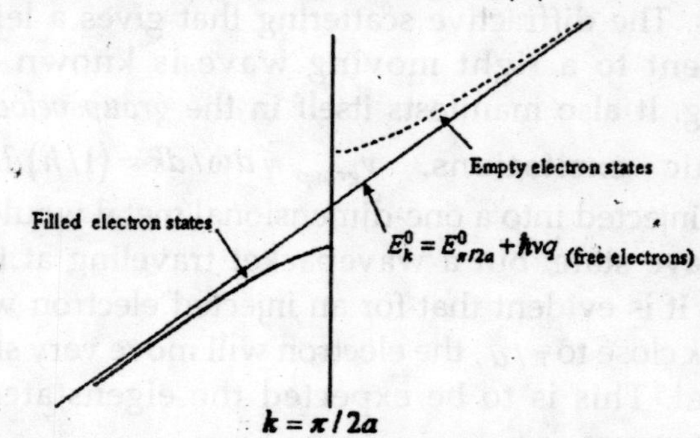

Change in electronic energy levels near $k = \pi/2a$ from dimerization: in this region, the free electron curve is approximated with a straight line: $q = k - \pi/2a$.

The crucial point is that if the atoms move together slightly into pairs, *the crystal has a new period* 2*a instead of a*. This means that the *potential* now has a nonzero component at $k = -\pi / a$, with a nonzero matrix element between the states

$k = \pi/2a$ and $k = -\pi/2a$, and so on. From this point, we can rerun the analysis above, except that now the gaps open up at $|k| = \pi/2a$ instead of at $|k| = \pi/a$.

The important point is that if the electrons fill all the states to $|k| = \pi/2a$, and none beyond (as would be the case for monovalent atoms) then the opening of a gap at $|k| = \pi/2a$ means that *all the electrons are in states whose energy is lowered.* To find the *total* energy benefit we need to integrate over *k*.

CALCULATING THE ELECTRONIC ENERGY GAINED BY DOUBLING THE LATTICE PERIOD

It is evident from the above that most of the contribution comes from fairly close to $k = \pi/2a$ (and of course symmetrically $k = -\pi/2a$). Since we want to find the total lowering in energy, let us study first the *bare* energy as a function of *k*, that is, the energy with no potential present. Of course, there isn't much to say: $E_k^0 = \hbar^2 k^2 / 2m$. However, the physics of these one-dimensional systems concerns only excitations near the "Fermi surface", the boundary between filled (low energy) states at zero temperature and empty states. This "Fermi surface" is in fact just two points in one dimension: $k = \pm\pi/2a$. In the neighbourhood of these two Fermi points, it is an excellent approximation to replace the gently curving $E_k^0 = \hbar^2 k^2 / 2m$ by *straight line approximations* the slope being $dE/dk = \hbar^2 k/m = \hbar p/m = \hbar v$.

Linearizing in the neighbourhood of $k = \pi/2a$, then, we take

$$E_k^0 = E_{\pi/2a}^0 + \hbar v(k - \pi/2a) = E_{\pi/2a}^0 + \hbar vq$$

where

$$q = k - \pi/2a$$

just *k* measured from the Fermi point $\pi/2a$. The variable *q* is *negative* for the relevant states, since they are on the lower energy side. The density of states in k-space is a constant $2\times L/2\pi = L/\pi$, remembering the two spin states per k-value. Recall

$$E_{\pm} = \frac{1}{2}(E_k^0 + E_{k-K}^0) \pm \sqrt{\left(\frac{E_k^0 - E_{k-K}^0}{2}\right)^2 + |V_K|^2}$$

but now $k = \pi/a$ and the lowering of energy of the electrons (counting it as a positive quantity) is:

$$2\int_0^{\pi/2a}\left(E_k^0 - E_-\right)Ldk/\pi = 2\int\left(\frac{1}{2}(E_k^0 - E_{k-K}^0) + \sqrt{\left(\frac{E_k^0 - E_{k-K}^0}{2}\right)^2 + |V_K|^2}\right)Ldk/\pi$$

where the extra factor of 2 counts the symmetrical contribution from the left-hand gap. (In examining the above expression, recall that for the $k > 0$ states we are interested in, $k > \pi/2a, E_k^0 - E_{k-k}^0$ is *negative*. The integrand on the right-hand side is still positive, very small for small k, reaching a maximum of $|V_k|$ at $k = \pi/2a$)

Putting in our linearized energy approximation,

$$E_k^0 = E_{\pi/2a}^0 + \hbar v\left(k - \pi/2a\right) = E_{\pi/2a}^0 + \hbar vq,$$

and remembering that now

$$E_{k-K}^0 = E_{-\pi/2a}^0 - \hbar v\left(k + \pi/2a\right) = E_{-\pi/2a}^0 - \hbar vq.$$

Since $E_{x/2a}^0 = E_{-x/2a}^0$,

$$\left(E_k^0 - E_{k-k}^0\right) = 2v\hbar q$$

Substituting these linearized values in the integral for the total energy lowering:

$$2\int_0^{\pi/2a}\left(E_k^0 - E_-\right)Ldk/\pi = 2\int_{-D}^{0}\left(v\hbar q + \sqrt{(v\hbar q)^2 + |V_K|^2}\right)Ldq/\pi$$

where in terms of the variable q we have set the lower limit of integration at $-D$: we can safely be vague about this lower limit, as the integral turns out to be logarithmic.

Since the integral is over negative numbers, and we have taken the positive square root, it is zero for zero $V_{K'}$ as it must be.

The integral can be done exactly, but it is more illuminating to divide the range of integration into $|v\hbar q| \le |V_k|$ and $|v\hbar q| > |V_k|$, then estimate the contributions from these two ranges separately.

First, consider $|v\hbar q| \le |V_k|$. Here the integrand is of order $|V_k|$, and the region Δq of integration corresponding to $|v\hbar q| \le |V_k|$ is of order so the integral over this range is of order.

Second, in the region $|v\hbar q| \le |V_k|$, we can write

$$2\int\left(v\hbar q + \sqrt{(v\hbar q)^2 + |V_K|^2}\right) Ldq/\pi = 2\int\left(v\hbar q + |v\hbar q|\sqrt{1 + \frac{|V_K|^2}{(v\hbar q)^2}}\right) Ldq/\pi$$

and expand the square root term. The leading terms cancel since q is negative, and the main contribution comes from the next term. This gives:

$$\Delta E \approx 2|V_K|^2 \int_{-D}^{-|V_K|} \frac{1}{2\hbar v} \frac{Ldq}{\pi|q|} = \frac{L|V_K|^2}{\hbar v} \ln \frac{|V_K|}{D}.$$

The important thing here is the logarithm. For sufficiently small $|V_k|$, this large (negative) term will dominate any term which is just proportional to V_k^2. But the elastic energy cost of the lattice "dimerizing" the atoms forming pairs, so that the distance between atoms alternates on going along the chain must be proportional to V_k^2. This leads to the conclusion that some, probably small, dimerization is always going to happen *a one-dimensional equally spaced chain with one electron per ion is unstable.*

Chapter 7

Van der Waals Forces Between Atoms

INTRODUCTION

The perfect gas equation of state $PV = NkT$ is manifestly incapable of describing actual gases at low temperatures, since they undergo a discontinuous change of volume and become liquids. In the 1870's, the Dutch physicist Van der Waals came up with an improvement: a gas law that recognized the molecules interacted with each other. He put in two parameters to mimic this interaction. The first, an attractive intermolecular force at long distances, helps draw the gas together and therefore reduces the necessary outside pressure to contain the gas in a given volume the gas is a little thinner near the walls. The attractive long range force can be represented by a negative potential $-aN/V$ on going away from the walls the molecules near the walls are attracted inwards, those in the bulk are attracted equally in all directions, so effectively the long range attraction is equivalent to a potential well extending throughout the volume, ending close to the walls. Consequently, the gas density N/V near the walls is decreased by a factor $e^{-E/kT} = e^{-aN/VkT} \cong 1 - aN/VkT$. Therefore, the pressure measured at the containing wall is from slightly diluted gas, so $P = (N/V)kT$ becomes $P = (N/V)\ (1 - aN/VkT)^{kT}$ or $\left(P + a(N/V)^2\right)V = NkT$. The second parameter van der Waals added was to take account of the finite molecular volume. A real gas cannot be compressed indefinitely it becomes a liquid, for all practical purposes incompressible. He represented this by replacing the volume V with $V - Nb$, Nb is referred to as the "excluded volume", roughly speaking the volume of the molecules, to give his famous equation

$$\left[P + a\left(\frac{N}{V}\right)^2\right](V - Nb) = NkT$$

This rather crude approximation does in fact give sets of isotherms representing the basic physics of a phase transition quite well. (For further details, and an enlightening discussion, see for example Appendix D of *Thermal Physics*, by R. Baierlein.)

GROUND STATE HYDROGEN ATOMS

Our interest here is in understanding the van der Waals long-range attractive force between electrically neutral atoms and molecules in quantum mechanical terms.

We begin with the simplest possible example, two hydrogen atoms, both in the ground state:

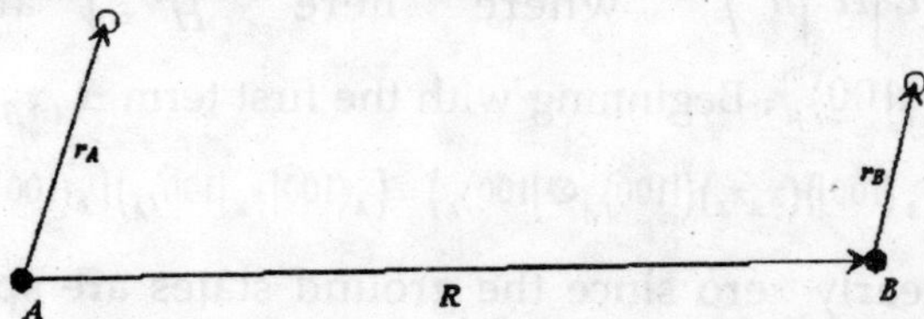

We label the atoms A and B, the vectors from the protons to the electron position are denoted by $\vec{r}_A$ and $\vec{r}_B$ respectively, and $\vec{R}$ is the vector from proton A to proton B.

Then the Hamiltonian $H = H^0 + V$, where

$$H^0 = -\frac{\hbar^2}{2m}\left(\nabla_A^2 + \nabla_B^2\right) - \frac{e^2}{r_A} - \frac{e^2}{r_B}$$

and the electrostatic interaction between the two atoms

$$V = \frac{e^2}{R} + \frac{e^2}{\left|\vec{R} + \vec{r}_B - \vec{r}_A\right|} - \frac{e^2}{\left|\vec{R} + \vec{r}_B\right|} - \frac{e^2}{\left|\vec{R} - \vec{r}_A\right|}.$$

The ground state of H^0 is just the product of the ground states of the atoms A, B, that is,

$$|0\rangle = |100\rangle_A \otimes |100\rangle_B .$$

Assuming now that the distance between the two atoms is much greater than their size, we can expand the interaction V in the small parameters r_A/R, r_B/R. As one might suspect from the diagram above, the leading order terms in the electrostatic energy are just those of a dipole-dipole interaction:

$$V = -e^2\left(\vec{r}_A\cdot\vec{\nabla}\right)\left(\vec{r}_B\cdot\vec{\nabla}\right)\frac{1}{R} = e^2\left[\frac{\vec{r}_A\cdot\vec{r}_B}{R^3} - \frac{3\left(\vec{r}_A\cdot\vec{R}\right)\left(\vec{r}_B\cdot\vec{R}\right)}{R^5}\right]$$

Taking now the z-axis in the direction, this interaction energy is

$$V = \frac{e^2}{R^3}\left(x_A x_B + y_A y_B - 2z_A z_B\right)$$

Now the first-order correction to the ground state energy of the two-atom system from this interaction is $E_n^1 = \left\langle n^0 \middle| H^1 \middle| n^0 \right\rangle$, where here $H^1 = V$ and $|n^0\rangle = |100\rangle_A \otimes |100\rangle_B$. Beginning with the first term $x_A x_B$ in V

$$\left({}_A\langle 100| \otimes {}_B\langle 100|\right)\left(x_A x_B\right)\left(|100\rangle_A \otimes |100\rangle_B\right) = \left({}_A\langle 100|x_A|100\rangle_A\right)\left({}_B\langle 100|x_B|100\rangle_B\right)$$

is clearly zero since the ground states are spherically symmetric. Similarly, the other terms in V are zero to first order.

Recall that the second-order energy correction is

$$E_n^2 = \sum_{m \neq n} \frac{\left|\left\langle m^0 \middle| H^1 \middle| n^0 \right\rangle\right|^2}{E_n^0 - E_m^0}.$$

That is,

$$E^{(2)} = \sum_{\substack{n,l,m \\ n',l',m'}} \frac{\left|\left({}_A\langle nlm| \otimes {}_B\langle n'l'm'|\right) V \left(|100\rangle_A \otimes |100\rangle_B\right)\right|^2}{2E_1 - E_n - E_{n'}}$$

A typical term here is

$$\left({}_A\langle nlm| \otimes {}_B\langle n'l'm'|\right)\left(x_A x_B\right)\left(|100\rangle_A \otimes |100\rangle_B\right)$$

$$= \left({}_A\langle nlm|x_A|100\rangle_A\right)\left({}_B\langle n'l'm'|x_B|100\rangle_B\right)$$

so the single-atom matrix elements are exactly those we discussed for the Stark effect (as we would expect this is an electrostatic interaction!). As before, only $l = 1$, $l' = 1$ contribute. To make a rough estimate of the size of $E^{(2)}$, we can use the same trick used for the quadratic Stark effect: replace the denominators by the constant $2E_1$ (the other terms are a lot smaller for the bound states, and continuum states have small overlap terms in the numerator). The sum over intermediate states n, l, m, n', l', m' can then be taken to be completely unrestricted, including even the ground state, giving

$$\sum_{\substack{n,l,m\\ n',l',m'}} \left(|nlm\rangle_A \otimes |n'l'm'\rangle_B\right)\left({}_A\langle nlm| \otimes_B \langle n'l'm'|\right) = I$$

the identity operator. In this approximation, then, just as for the Stark effect,

$$E^{(2)} = \frac{e^4}{R^6}\frac{1}{2E_1}\left({}_A\langle 100| \otimes_B \langle 100|\right)\left(x_A x_B + y_A y_B - 2z_A z_B\right)^2\left(|100\rangle_A \otimes |100\rangle_B\right)$$

where $E_1 = -1$ Ryd., so this is a *lowering* of energy.

In multiplying out $\left(x_A x_B + y_A y_B - 2z_A z_B\right)^2$, the cross terms will have expectation values of zero. The ground state wave function is symmetrical, so all we need is $\left\langle 100\left|x^2\right|100\right\rangle = a_0^2$, where a_0 is the Bohr radius.

This gives

$$E^{(2)} = \frac{e^4}{R^6}\frac{1}{2E_1}6a_0^4 = -6\frac{e^2}{R}\left(\frac{a_0}{R}\right)^5$$

using $E_1 = -e^2/2a_0$. Bear in mind that this is an approximation, but a pretty good one a more accurate calculation replaces the 6 by 6.5.

FORCES BETWEEN A 1*S* HYDROGEN ATOM AND A 2*P* HYDROGEN ATOM

With one atom in the $|100\rangle$ and the other in $|210\rangle$, say, a typical leading order term would be

$$\left({}_A\langle 100|\otimes{}_B\langle 210|\right)(x_A x_B)\left(|100\rangle_A\otimes|100\rangle_B\right)=\left({}_A\langle 100|x_A|100\rangle_A\right)\left({}_B\langle 210|x_B|100\rangle_B\right),$$

and this is certainly zero, as are all the other leading terms. Baym (*Lectures on Quantum Mechanics*) concluded from this that there is *no* leading order energy correction between two hydrogen atoms if one of them is in the ground state. This is incorrect: the first excited state of the two-atom system (without interaction) is degenerate, so, exactly as for the 2-D simple harmonic oscillator treated in the previous lecture, we must diagonalize the perturbation in the subspace of these degenerate first excited states.

The space of the degenerate first excited states of the two noninteracting atoms is spanned by the product-space kets:

$$\left(|100\rangle_A\otimes|200\rangle_B\right),\ \left(|200\rangle_A\otimes|100\rangle_B\right),\ \left(|100\rangle_A\otimes|211\rangle_B\right),\ \left(|211\rangle_A\otimes|100\rangle_B\right),$$
$$\left(|100\rangle_A\otimes|210\rangle_B\right),\ \left(|210\rangle_A\otimes|100\rangle_B\right),\ \left(|100\rangle_A\otimes|21-1\rangle_B\right),\ \left(|21-1\rangle_A\otimes|100\rangle_B\right).$$

The task, then, is to diagonalize $V=\frac{e^2}{R^3}$ $\left(x_A x_B + y_A y_B - 2z_A z_B\right)$ in this eight-dimensional subspace.

We begin by representing V as an 8‰8 matrix using these states as the basis. First, note that all the diagonal elements of the matrix are zero. Second, writing $V=\frac{e^2}{R^3}\left(\vec{r}_A\cdot\vec{r}_B-3z_A z_B\right)$, it is evident that V is unchanged if the system is rotated around the z-axis (the line joining the two protons). This means that the commutator $[V,L_x]=0$, where L_z is the total angular momentum component in the z-direction, so V will only have nonzero matrix elements between states having the same total. Third, from parity (or Wigner-Eckart) all matrix elements in the subspace spanned by $\left(|100\rangle_A\otimes|200\rangle_B\right)$, $\left(|200\rangle_A\otimes|100\rangle_B\right)$ must be zero.

This reduces the nonzero part of the 8‰8 matrix to a direct product of three 2‰2 matrices, corresponding to the three values of $= m$. For example, the $m = 0$ subspace is spanned by. The diagonal elements of the 2‰2 matrix are zero, the off-diagonal elements are equal to $-\frac{2e^3}{R^3}\left({}_A\langle 100|z_A|210\rangle_A\right)$ $\left({}_B\langle 210|z_B|100\rangle_B\right)$, where we have kept the unnecessary labels *A*, *B* to make clear where this term comes from. (The x_A and y_A terms will not contribute for $m = 0$.)

This is now a straightforward integral over hydrogen wave functions. The three 2‰2 matrices have the form

$$\begin{pmatrix} 0 & k_m / R^3 \\ k_m / R^3 & 0 \end{pmatrix}$$

(following the notation of Cohen-Tannoudji) where $k_m \sim e^2 a_0^2$, and the energy eigenvalues are $\pm k_m / R^3$, with corresponding eigenkets.

$$(1/\sqrt{2})\left[\left(|100\rangle_A \otimes |210\rangle_B\right) \pm \left(|210\rangle_A \otimes |100\rangle_B\right)\right]$$

So for two hydrogen atoms, one in the ground state and one in the first excited state, the van der Waal interaction energy goes as $1/R^3$, much more important than the energy $1/R^6$ for two hydrogen atoms in the ground state. Notice also that the $1/R^3$ can be *positive or negative,* depending on whether the atoms are in an even or an odd state so the atoms sometimes repel each other.

Finally, if two atoms are initially in a state $\left(|100\rangle_A \otimes |210\rangle_B\right)$, note that this is *not* an eigenstate of the Hamiltonian when the interaction is included. Writing the state as a sum of the even and odd states, which have slightly different phase frequencies from the energy difference, we find the excitation moves back and forth between the two atoms with a period $hR^3 / 2k_{m-0}$.

Chapter 8

Higher Order Perturbation Theory

THE INTERACTION REPRESENTATION

Recall that in the first part of this course sequence, we discussed the Schrödinger and Heisenberg representations of quantum mechanics here. In the Schrödinger representation, the operators are time-independent (except for explicitly time-dependent potentials) the kets representing the quantum states develop in time. In the Heisenberg representation, the kets stay the same, the time dependence is in the operators. These differing representations describe the same physics matrix elements of operators between kets must be the same in both. The most natural to use depends on the problem at hand. In the classical limit, for example, the Heisenberg operators have the time dependence of the corresponding classical operators.

In fact, for perturbation theory problems with a time-dependent potential, an intermediate representation, the *interaction representation,* is very convenient. Using a subscript S to denote the Schrödinger representation,

$$i\hbar\frac{d}{dt}\left|\psi_S(t)\right\rangle = H_S\left|\psi_S(t)\right\rangle = \left(H_S{}^0 + V_S(t)\right)\left|\psi_S(t)\right\rangle,$$

we define the *interaction representation* by the unitary transformation

$$\left|\psi_I(t)\right\rangle = e^{iH_S^0 t/\hbar}\left|\psi_S(t)\right\rangle$$

so the interaction representation kets and the Schrödinger representation kets coincide at $t = 0$, and if the interaction were zero, the interaction representation kets would be constant in time, like those in the Heisenberg representation.

For nonzero v(t), then, the time development of the interaction representation kets is entirely due to v(t), and is easily found by differentiating both sides of the equation:

$$i\hbar \frac{d}{dt} | \psi_1(t) \rangle = -H^0 | \psi_1(t) \rangle + e^{iH_S^0 t/\hbar} \frac{d}{dt} | \psi_s(t) \rangle$$

$$= = -H^0 | \psi_1(t) \rangle + e^{iH_S^0 t/\hbar} \left(H_S^0 + V_s(t) \right) e^{-iH_S^0 t/\hbar} | \psi_1(t) \rangle$$

$$= e^{iH_S^0 t/\hbar} V_S(t) e^{-iH_S^0 t/\hbar} | \psi_1(t) \rangle$$

where we have introduced the interaction representation operator $V_I(t)$, defined by

$$V_1(t) = e^{iH_s^0 t/\hbar} V_s(t) e^{-iH_s^0 t/\hbar}$$

Operators in this representation must have this time dependence relative to the Schrödinger operators to ensure that matrix elements, the only quantities of physical significance, are the same in the two representations. That is to say, we must have

$$\langle f_I^0 | O_I | i_I^0 \rangle = \langle f_S^0 | O_S | i_S^0 \rangle,$$

the two representations must predict the same probability amplitude for any transition.

Integrating both sides of the differential equation,

$$|\psi_I(t)\rangle = |\psi_I(0)\rangle - \frac{i}{\hbar} \int_0^t dt' V_I(t') |\psi_I(t')\rangle.$$

This is not a solution we've just gone from a differential equation to an integral equation. This is only worth doing if V_I is small, in which case the integral equation can be solved iteratively.

The zeroth approximation is then

$$|\psi_I(t)\rangle = |\psi_I(0)\rangle.$$

Putting this value into the small term on the right hand side of the integral equation gives the first order solution,

$$|\psi_I(t)\rangle = |\psi_I(t_0)\rangle - \frac{i}{\hbar}\int_0^t dt' V_I(t')|\psi_I(0)\rangle.$$

The second order solution is now given by putting the first order solution into the integral on the right:

$$|\psi_I(t)\rangle = |\psi_I(0)\rangle - \frac{i}{\hbar}\int_0^t dt' V_I(t')\left(|\psi_I(0)\rangle - \frac{i}{\hbar}\int_0^{t'} dt'' V_I(t'')|\psi_I(0)\rangle\right).$$

This can be written:

$$|\psi_I(t)\rangle = \left(1 - \frac{i}{\hbar}\int_0^t dt' V_I(t') + \left(-\frac{i}{\hbar}\right)^2 \int_0^t dt' \int_0^{t'} dt'' V_I(t') V_I(t'')\right)|\psi_I(0)\rangle.$$

The complete perturbation series is generated by repeating the iteration to all orders. It can be expressed as a *time-ordered product*:

$$|\psi_I(t)\rangle = T\exp\left(-\frac{i}{\hbar}\int_0^t dt' V_I(t')\right)|\psi_I(0)\rangle.$$

The T symbol means that on expanding out the exponential, the operators at different times are arranged in order of time, the latest on the left, without worrying about commutators. If we just blindly expand the exponential, we will get, for example, a third-order term

$$T\frac{1}{3!}\left(-\frac{i}{\hbar}\int_0^t dt' V_I(t')\right)\left(-\frac{i}{\hbar}\int_0^t dt'' V_I(t'')\right)\left(-\frac{i}{\hbar}\int_0^t dt''' V_I(t''')\right).$$

The T operator tells us to rearrange the $V_I(t)$'s in chronological order. Since there are three of them, they clearly appear in all possible orders before T operates, that is to say, there are 3! different ordered terms that T makes the same. This just nicely cancels the 3! in the exponential expansion, to give us the expression we found by iteration.

This time-ordered exponential is therefore the *interaction representation propagator*:

$$|\psi_I(t)\rangle = U_I(t,0)|\psi_I(0)\rangle, \quad U_I(t,0) = T\exp\left(-\frac{i}{\hbar}\int_0^t dt' V_I(t')\right).$$

GOING BACK TO THE SCHRÖDINGER REPRESENTATION

It is instructive to recast this result in the Schrödinger representation. First, note that putting the above equation for U_I together with the original definition of interaction representation kets

$$\left|\psi_1(t)\right\rangle = e^{iH_s^0 t/\hbar}\left|\psi_S(t)\right\rangle$$

gives $|\psi_S(t)\rangle = e^{-iH_S^0 t/\hbar}|\psi_1(t)\rangle = e^{-iH_S^0 t/\hbar}U_I(t,0)|\psi_1(0)\rangle$

$$= e^{-iH_S^0 t/\hbar}U_I(t,0)|\psi_S(0)\rangle$$

So the Schrödinger representation propagator is related to the interaction representation propagator by:

$$U_S(t,0) = e^{-iH_s^0 t/\hbar}U_1(t,0)$$

Now let us see how to put our perturbation expansion for the propagator back from the interaction representation into the Schrödinger representation. Instead of trying to handle the whole infinite series at once, we concentrate on the second-order term. We will discover a pattern that works for all the higher order terms as well.

So, transforming the operators in the second-order term of the interaction propagator back to the Schrödinger form,

using $V_1(t) = e^{iH_s^0 t/\hbar}V_S(t)e^{-iH_s^0 t/\hbar}$

we find

$$\left(\frac{1}{i\hbar}\right)^2 \int_0^t dt' \int_0^r dt'' V_1(t'') = \left(\frac{1}{i\hbar}\right)^2 \int_0^t\int_0^r dt' dt'' e^{iH_S^0/\hbar} e^{iH_S^0 r/\hbar V_S(t'') e^{-iH_S^0 ''/\hbar}}$$

Recall also that the Schrödinger propagator has the extra term multiplying the interaction representation propagator. Putting this in, and combining some of the exponentials, we find the second-order contribution to the Schrödinger propagator to be:

$$\left(\frac{1}{i\hbar}\right)^2 \int_0^t \int_0^{t'} dt'dt'' e^{-H_S^0(t-t'')/\hbar} V_S(t') e^{-iH_S^0(t'-t'')/\hbar} V_S(t'') e^{-iH_S^0 t''/\hbar}$$

which can also be written:

$$\left(\frac{1}{i\hbar}\right)^2 \int_0^t \int_0^{t'} dt'dt'' U_S^0(t,t') V_S(t') U_S^0(t',t'') V_S(t'') U_S^0(t'',0).$$

To find the probability amplitude corresponding to this second-order process, we must sandwich it between initial and final states. We take as our basis set the eigenstates of $H_S{}^0$. If we insert the unit operator

$$I = \sum_n \left|n^0\right\rangle\left\langle n^0\right|$$

between the two V_S's, the exponentiated $H_S{}^0$'s become simply numbers since they are now acting on eigenstates, and the expression becomes

$$\left(\frac{1}{i\hbar}\right)^2 \int_0^t \int_0^{t'} dt'dt'' e^{-iE_f^0(t-t')/\hbar} \left\langle f^0\middle|V_S(t')\middle|n^0\right\rangle$$

$$e^{-iE_n^0(t'-t'')/\hbar} \left\langle n^0\middle|V_S(t'')\middle|i^0\right\rangle e^{-iE_i^0 t''/\hbar}$$

The interpretation is now clear: the initial state $\left|i^0\right\rangle$ evolves from $t = 0$ to t'' under $H_S{}^0$, that is to say, only its phase changes in the standard fashion. At t'' the interaction $V_S(t'')$ kicks it into another eigenstate $\left|n^0\right\rangle$ of $H_S{}^0$, and only the phase changes until t', when $V_S(t')$ sends it to the final state$|f\rangle$. This process must be summed over all times t', t'' between t_0 and t, and over all possible intermediate states.

The nth order term has precisely the same structure, with V_S coming into play n times.

Chapter 9

Time-Dependent Perturbation Theory

INTRODUCTION: GENERAL FORMALISM

We look at a Hamiltonian $H = H^0 + V(t)$, with $v(t)$ some time-dependent perturbation, so now the wave function will have perturbation-induced time dependence.

Our starting point is the set of eigenstates $|n\rangle$ of the unperturbed Hamiltonian $H^0|n\rangle = E_n|n\rangle$, notice we are not labeling with a zero, no E_n^0, because with a time-dependent Hamiltonian, energy will not be conserved, so it is pointless to look for energy corrections. What happens instead, provided the perturbation is not too large, is that the system makes transitions between the eigenstates $|n\rangle$ of H^0.

Of course, even for $V = 0$, the wave functions have the usual time dependence,

$$|\psi(t)\rangle = \sum_n c_n e^{-E_n t/\hbar}|n\rangle$$

with the c_n's constant. What happens on introducing $v(t)$ is that the c_n's *themselves* acquire time dependence,

$$|\psi(t)\rangle = \sum_n c_n(t) e^{-E_n t/\hbar}|n\rangle$$

and this time dependence is determined by Schrödinger's equation with $H = H^0 + V(t)$:

$$i\hbar\frac{\partial}{\partial t}\sum_n c_n(t)e^{-iE_nt/\hbar}|n\rangle = \left(H^0+V(t)\right)\sum_n c_n(t)e^{-iE_nt/\hbar}|n\rangle$$

so $$i\hbar\sum_n \dot{c}_n(t)e^{-iE_nt/\hbar}|n\rangle = V(t)\sum_n c_n(t)e^{-iE_nt/\hbar}|n\rangle$$

Taking the inner product with the bra $\langle m|e^{iE_mt/\hbar}$, and introducing $\omega_{mn}=\frac{E_m-E_n}{\hbar}$,

$$i\hbar c_m = \sum_n \langle m|V(t)|n\rangle c_n e^{i\omega_{mn}t} = \sum_n V_{mn}e^{i\omega_{mn}t}c_n$$

This is a matrix differential equation for the c_n's:

$$i\hbar\begin{pmatrix}\dot{c}_1\\ \dot{c}_2\\ \dot{c}_3\\ \cdot\end{pmatrix} = \begin{pmatrix}V_{11} & V_{12}e^{i\omega_{12}t} & \cdot & \cdot & \cdot\\ V_{21}e^{i\omega_{21}t} & V_{22} & \cdot & \cdot & \cdot\\ \cdot & \cdot & V_{33} & \cdot & \cdot\\ \cdot & \cdot & \cdot & & \end{pmatrix}\begin{pmatrix}c_1\\ c_2\\ c_3\\ \cdot\end{pmatrix}$$

and solving this set of coupled equations will give us the $c_n(t)$'s, and hence the probability of finding the system in any particular state at any later time.

If the system is in initial state $|i\rangle$ at $t=0, |f\rangle$ the probability amplitude for it being in state at time t is *to leading order* in the perturbation

$$C_f(t)=\delta_{fi}-\frac{i}{\hbar}\int_0^t V_{fi}(t')e^{i\omega_{fi}t'}dt'$$

The probability that the system is in fact in state $|f\rangle$ at time t is therefore

$$|C_f(t)|^2=\frac{i}{\hbar}\left|\int_0^t V_{fi}(t')e^{i\omega_{fi}t'}dt'\right|^2$$

Obviously, this is only going to be a good approximation if it predicts that the probability of transition is small otherwise we need to go to higher order, using the Interaction Representation

Kicking an Oscillator

Suppose a simple harmonic oscillator is in its ground state $|0\rangle$ at $t = -\infty$. It is perturbed by a small time-dependent potential $V(t) = -eExe^{-t^2/\tau^2}$. What is the probability of finding it in the first excited state $|1\rangle$ at $t = +\infty$?

Here $V_{fi}(t') = -eE\langle 1|x|0\rangle e^{-t^2/x^2}$, and

$x = \sqrt{\hbar/2m\omega}\left(a + a^{\uparrow}\right)$, from which the probability can be evaluated. It is $\left(e^2E^2/\hbar^2\right)(\hbar/2m\omega)\pi\tau^2 e^{-\omega^2 x^2/2}$

It's worth thinking through the physical interpretations for very long and for very short times, and explaining the significance of the time for which the probability is a maximum.

THE TWO-STATE SYSTEM: AN EXACT SOLUTION

For the particular case of a two-state system perturbed by a periodic external field, the matrix equation above can be solved exactly. Of course, real physical systems have more than two states, but in fact for some important cases two of the states may be only weakly coupled to other degrees of freedom and the analysis then becomes relevant. A famous example, the ammonia maser, is discussed at the end of the section.

For a two-state system, then, the most general wave function is $|\psi(t)\rangle = c_1(t)e^{-iE_1t/\hbar}|1\rangle + c_2(t)e^{-iE_2t/\hbar}|2\rangle$

and the differential equation for the $c_n(t)$'s is:

$$i\hbar\begin{pmatrix}\dot{C}_2\\ \dot{C}_2\end{pmatrix} = \begin{pmatrix}0 & Ve^{i\omega t}e^{-i\omega_{12}t}\\ Ve^{-i\omega t}e^{-i\omega_{12}t} & 0\end{pmatrix}\begin{pmatrix}C_1\\ C_2\end{pmatrix}$$ Writing

$\omega + \omega_{12} = \alpha$ for convenience, the coupled equations are:

$$i\hbar\dot{c}_1 = Ve^{i\alpha t}c_2$$

$$i\hbar\dot{c}_2 = Ve^{-i\alpha t}c_1$$

These two first-order equations can be transformed into a single second-order equation by differentiating the second one, then substituting $\dot{C}_1$ from the first one and C_1 from the second one to give

$$\ddot{c}_2 = -i\alpha\dot{c}_2 - \frac{V^2}{\hbar^2}c_2.$$

This is a standard second-order differential equation, solved by putting in a trial solution $C_2(t) = C_2(0)e^{i\Omega t}$. This satisfies the equation if $\Omega = -\frac{\alpha}{2} \pm \sqrt{\frac{\alpha^2}{4} + \frac{V^2}{\hbar^2}}$, so, reverting to the original $\omega + \omega_{12} = \alpha$, the general solution is:

$$C_2(t) = e^{-i\frac{\left(w - w^{2l}\right)}{2}t}\left(Ae^{i\sqrt{\left(\frac{w - w_{2l}}{2}\right)^2 + \frac{V^2}{k^2}}t} + Be^{i\sqrt{\left(\frac{w - w_{2l}}{2}\right)^2 + \frac{V^2}{k^2}}t}\right).$$

Taking the initial state to be $c_1(0) = 1, c_2(0) = 0$ gives $A = -B$.

To fix the overall constant, note that at $t = 0$,

$\dot{c}_2(0) = \frac{V}{i\hbar}c_1(0) = \frac{V}{i\hbar}$. Therefore

$$|c_2(t)|^2 = \frac{\frac{V^2}{\hbar^2}}{\left(\frac{\omega - \omega_{21}}{2}\right)^2 + \frac{V^2}{\hbar^2}}\sin^2\left(\sqrt{\left(\frac{\omega - \omega_{21}}{2}\right)^2 + \frac{V^2}{\hbar^2}}\,t\right).$$

Note in particular the result if $\omega = \omega_{12}$: $|C_2(t)|^2 = \sin^2\left(\frac{Vt}{\hbar}\right)$. Assuming $E_2 > E_1$, and the two-state system to be initially in the ground state $|1\rangle$, this means that after a time $h/4V$ the system will *certainly* be in state$|2\rangle$, and will oscillate back and forth between the two states with period $h/2V$.

That is to say, a precisely timed period spent in an oscillating field can drive a collection of molecules all in the ground state to be all in an excited state. The ammonia *maser* works by sending a stream of ammonia molecules, traveling at known velocity, down a tube having an oscillating field for a definite length, so the molecules emerging at the other end are all (or almost all, depending on the precision of ingoing velocity, etc.) in the first excited state. Application of a small amount of electromagnetic radiation of the same frequency to the outgoing molecules will cause some to decay, generating intense radiation and therefore a much shorter period for all to decay, emitting coherent radiation.

A "SUDDEN" PERTURBATION

A sudden perturbation is defined here as a sudden switch from one time-independent Hamiltonian H_0 to another one $H_0^{'}$, the time of switching being much shorter than any natural period of the system. In this case, perturbation theory is irrelevant: if the system is initially in an eigenstate $|n\rangle$ of H_0, one simply has to write it as a sum over the eigenstates of $H_0^{'}$, $|n\rangle = \sum_{n'} |n'\rangle\langle n'|n\rangle$. The nontrivial part of the problem is in establishing that the change *is* sudden enough, by estimating the actual time taken for the Hamiltonian to change, and the periods of motion associated with the state $|n\rangle$ and with its transitions to neighbouring states.

(We discussed one example last semester an electron in the ground state in a one-dimensional box that suddenly doubles in size. Other favorite examples include an atom with spin-orbit coupling in a magnetic field that suddenly reverses (Messiah p 743), and the reaction of orbiting electrons to nuclear α- or β-decay.)

HARMONIC PERTURBATIONS: FERMI'S GOLDEN RULE

Let us consider a system in an initial state $|i\rangle$ perturbed by a periodic potential $V(t) = Ve^{-iwt}$ switched on at $t = 0$. For example, this could be an atom perturbed by an external oscillating electric field, such as an incident light wave.

What is the probability that at a later time t the system be in state $|f\rangle$?

Recall the matrix differential equation for the c_n's:

$$i\hbar \begin{pmatrix} \dot{c}_1 \\ \dot{c}_2 \\ \dot{c}_3 \\ \cdot \\ \cdot \end{pmatrix} = \begin{pmatrix} V_{11} & V_{12}e^{i\omega_{12}t} & \cdot & \cdot & \cdot \\ V_{21}e^{i\omega_{21}t} & V_{22} & \cdot & \cdot & \cdot \\ \cdot & \cdot & V_{33} & \cdot & \cdot \\ \cdot & \cdot & \cdot & \cdot & \cdot \\ \cdot & \cdot & \cdot & \cdot & \cdot \end{pmatrix} \begin{pmatrix} c_1 \\ c_2 \\ c_3 \\ \cdot \\ \cdot \end{pmatrix}$$

Since the system is definitely in state $|i\rangle$ at $t = 0$, the ket vector on the right is initially $C_i = 1, C_{j\neq i} = 0$.

The first-order approximation to keep the vector $C_i = 1, C_{j\neq i} = 0$ on the right, that is, to solve the equations

$$i\hbar\dot{c}_f(t) = V_{fi}e^{i\omega_f t}$$

Integrating this equation, the probability amplitude for an atom in initial state $|i\rangle$ to be in state $|f\rangle$ after time t is, to first order:

$$C_f(t) = -\frac{i}{\hbar}\int_0^t \langle f|V|i\rangle e^{i(w_{fi}-w)t'}$$

$$= -\frac{i}{\hbar}\int_0^t \langle f|V|i\rangle \frac{e^{i(w_{fi}-w)t} - 1}{i(w_{fi}-w)}$$

The probability of transition is therefore

$$P_{i\to f}(t)=|c_f|^2=\frac{1}{\hbar^2}|\langle f|V|i\rangle|^2\left(\frac{\sin((\omega_{fi}-\omega)t/2)}{(\omega_{fi}-\omega)/2}\right)^2$$

and we're interested in the large t limit.

Writing $\alpha=(\omega_{fi}-\omega)/2$, our function has the form $\dfrac{\sin^2\alpha t}{\alpha^2}$.

This function has a peak at $\alpha = 0$, with maximum value t^2, and width of order $1/t$, so a total weight of order t. The function has more peaks at $\alpha t = (n + 1/2)\pi$. These are bounded by the denominator at $1/\alpha^2$. For large t their contribution comes from a range of order $1/t$ also, and as $t\to\infty$ the function tends to a δ function at the origin, but multiplied by t.

This divergence is telling us that there is a finite probability *rate* for the transition, so the likelihood of transition is proportional to time elapsed. Therefore, we should divide by t to get the transition rate.

To get the quantitative result, we need to evaluate the weight of the function term. We use the standard result $\int_{-\infty}^{\infty}\left(\frac{\sin\xi}{\xi}\right)^2 d\xi=\pi$ to find $\int_{-\infty}^{\infty}\left(\frac{\sin\alpha t}{\alpha}\right)^2 d\alpha=\pi t$, and therefore

$$\lim_{t\to\infty}\frac{1}{t}\left(\frac{\sin\alpha t}{\alpha}\right)^2=\pi\delta(\alpha).$$

Now, the transition rate is the probability of transition divided by t in the large t limit, that is,

$$R_{i\to f}(t)=\lim_{t\to\infty}\frac{P_{i\to f}(t)}{t}=\lim_{t\to\infty}\frac{1}{t}\frac{1}{\hbar^2}|\langle f|V|i\rangle|^2\left[\frac{\sin((\omega_{fi}-\omega)t/2)}{(\omega_{fi}-\omega)/2}\right]$$

$$=\frac{1}{\hbar^2}|\langle f|V|i\rangle|^2\pi\delta\left(\tfrac{1}{2}(\omega_{fi}-\omega)\right)$$

$$=\frac{2\pi}{\hbar^2}|\langle f|V|i\rangle|^2\delta(\omega_{fi}-\omega)$$

This last line is Fermi's Golden Rule: we shall be using it a lot. You might worry that in the long time limit we have taken the probability of transition is in fact diverging, so how can we use first order perturbation theory? The point is that for a transition with $\omega_{fi} \neq \omega$, "long time" means $(\omega_{fi} - \omega)t >> 1$, this can still be a very short time compared with the mean transition time, which depends on the matrix element. In fact, Fermi's Rule agrees extremely well with experiment when applied to atomic systems.

ANOTHER DERIVATION OF THE GOLDEN RULE

Actually, when light falls on an atom, the full periodic potential is not suddenly switched on, on an atomic time scale, but builds up over many cycles (of the atom and of the light). Baym re-derives the Golden Rule assuming the limit of a very slow switch on, $V(t) = e^{it} Ve^{-i\omega t}$ with ε very small, so V switched on very gradually in the past, and we are looking at times much smaller than $1/\varepsilon$. We can then take the initial time to be, that is,

$$C_f(t) = -\frac{i}{\hbar}\int_{-\infty}^{t} \langle f|V|i\rangle e^{i(w_{fi}-w-i\tau)t'} dt' = -\frac{1}{\hbar}\frac{e^{i(w_{fi}-w-i\tau)t}}{w_{fi}-w-i\varepsilon}\langle f|V|i\rangle$$

so

$$|c_f(t)|^2 = \frac{1}{\hbar^2}\frac{e^{2\varepsilon t}}{(\omega_{fi}-\omega)^2+\varepsilon^2}|\langle f|V|i\rangle|^2$$

and the time rate of change

$$\frac{d}{dt}|C_f(t)^2| = \frac{1}{\hbar^2}\frac{2\varepsilon e^{2\tau t}}{(\omega_{fi}-\omega)^2+\varepsilon^2}|\langle f|V|i\rangle|^2.$$

In the limit $\varepsilon \to 0$, the function

$$\frac{2\varepsilon}{(\omega_{fi}-\omega)^2+\varepsilon^2} \to 2\pi\delta(\omega_{fi}-\omega)$$

giving the Golden Rule again.

HARMONIC PERTURBATIONS: SECOND-ORDER TRANSITIONS

Sometimes the first order matrix element $\langle f|V|i\rangle$ is identically zero (parity, Wigner Eckart, etc.) but other matrix elements are nonzero and the transition can be accomplished by an indirect route. In the notes on the interaction representation, we derived the probability amplitude for the second-order process,

$$C_n^{(2)}(t)=\left(\frac{1}{i\hbar}\right)^2$$

$$=\sum_n \int_0^t \int_0^r dt'd''e^{-iw\tau(t-t')} <f|V_s(t')|n> e^{-w_n(t'-t'')} <n|V_s(t'')|i> e^{iw_i r}$$

(writing $E_n/\hbar=\omega_n$, etc.)Taking the gradually switched-on harmonic perturbation $V_s(t)=e^{\tau t}Ve^{-i\omega t}$, and the initial time $-\infty$, as above,

$$C_n^{(2)}(t)=\left(\frac{1}{i\hbar}\right)^2 \sum_n <f|V|n><n|V|i> e^{-i\omega\tau t}$$

$$\int_{-\infty}^{t} dt' \int_{-\infty}^{r} dt'' e^{i(w\tau-w_n-w-i\tau)t'} e^{i(w_n-w_i-w-i\tau)t''}$$

The integrals are straightforward, and yield

$$c_n^{(2)}(t)=\left(\frac{1}{i\hbar}\right)^2 e^{-i(\omega_i-\omega_f)t} \frac{e^{2\varepsilon t}}{\omega_f-\omega_i-2\omega-2i\varepsilon}\sum_n \frac{<f|V|n><n|V|i>}{\omega_n-\omega_i-\omega-i\varepsilon}.$$

Exactly as in the section above on the first-order Golden Rule, we can find the transition rate:

$$\frac{d}{dt}\left|c_n^{(2)}(t)\right|^2=\frac{2\pi}{\hbar^4}\left|\sum_n \frac{<f|V|n><n|V|i>}{\omega_n-\omega_i-\omega-i\varepsilon}\right|^2 \delta(\omega_f-\omega_i-2\omega).$$

(The $\hbar^4$ in the denominator goes to $\hbar$ on replacing the frequencies ω with energies E, both in the denominator and the delta function, remember that if $E = \hbar\omega, \delta(\omega) = \hbar\delta(E)$)

This is a transition in which the system gains energy $2\hbar\omega$ from the beam, in other words *two* photons are absorbed, the first taking the system to the intermediate energy ω_n, which is short-lived and therefore not well defined in energy there is no energy conservation requirement into this state, only between initial and final states.

Of course, if an atom in an arbitrary state is exposed to monochromatic light, other second order processes in which two photons are emitted, or one is absorbed and one emitted (in either order) are also possible.

Chapter 10

Photoelectric Effect in Hydrogen

INTRODUCTION

In the photoelectric effect, incoming light causes an atom to eject an electron. We consider the simplest possible scenario: that the atom is hydrogen in its ground state. The interesting question is: for an ingoing light wave of definite frequency and amplitude, what is the probability of ionization of a hydrogen atom in a given time? In other words, assuming we can use time-dependent perturbation theory, what is the ionization rate?

Formally, we know what to do. We must find the interaction Hamiltonian, then use Fermi's Golden Rule for the transition rate with a periodic perturbation:

$$R_{i\to f} = \frac{2\pi}{\hbar}\left|\langle f|H^1|i\rangle\right|^2 \delta\left(E_f - E_i - \hbar\omega\right)$$

But it's not that easy! For one thing, the outgoing electron will be in some kind of plane wave state, so whatever convention we adopt for normalizing such states appears in the rate. But also the δ function is tricky for excitation into the continuum: just how many of these plane wave states satisfy $E_f = E_i + \hbar\omega$? We shall discover that with a consistent formalism, these two difficulties cancel each other.

THE INTERACTION HAMILTONIAN

Taking the incoming wave to be an electromagnetic field having vector potential.

$$\vec{A}(\vec{r},t) = \vec{A}_0 \cos\left(\vec{k}\cdot\vec{r} - \omega t\right)$$

The interaction Hamiltonian is given by replacing the electron kinetic energy term $\vec{p}^2/2m$ with $\left(\vec{p} - q\vec{A}/c\right)^2/2m$. The relevant new term is

$$-(1/2m)\left(q/c\left(\vec{p}\cdot\vec{A} + \vec{A}\cdot\vec{p}\right)\right) = (e/mc)\vec{A}\cdot\vec{p}$$

since $q = -e$ and $\vec{\nabla}\cdot\vec{A} = 0$ in our gauge.

Therefore

$$H^1 = \left(\frac{e}{mc}\right)\cos\left(\vec{k}\cdot\vec{r} - \omega t\right)\vec{A}_0\cdot\vec{p}$$

$$= \left(\frac{e}{2mc}\right)\left(e^{i(\vec{k}\cdot\vec{r}-\omega t)} + e^{-i(\vec{k}\cdot\vec{r}-\omega t)}\right)\vec{A}_0\cdot\vec{p}.$$

The two different terms in this expression, having time dependences $e^{-i\omega t}$ and $e^{i\omega t}$ will give δ functions $\delta\left(E_f - E_i - \hbar\omega\right)$ and $\delta\left(E_f - E_i + \hbar\omega\right)$ respectively in the transition rate. The $e^{-i\omega t}$ term therefore corresponds to absorption of a photon, since we are looking at a process in which the electron gains energy, $E_f > E_i$. The $e^{i\omega t}$ term is for the process where an atom in an excited state emits a photon into the beam and *drops* in energy.

So the relevant interaction Hamiltonian is

$$H^1(t) = H^1 e^{-i\omega t} \text{ where } H^1 = \left(\frac{e}{2mc}\right)e^{i\vec{k}\cdot\vec{r}}\vec{A}_0\cdot\vec{p}.$$

PLANE WAVES: DENSITY OF STATES

We make the assumption that the final state is a plane wave state $|\vec{k}_f\rangle \alpha e^{i\vec{k}_f\cdot\vec{r}}$.

The most straightforward way of handling the plane wave states is to confine the whole system to an extremely large cubical box of side L, and impose *periodic* boundary conditions (so that plane traveling wave states are allowed).

The big box has volume V = L^3, so the appropriately normalized plane wave states are

$$|\vec{k}\rangle = \frac{1}{L^{3/2}} e^{i\vec{k}.\vec{r}} = \frac{1}{\sqrt{V}} e^{i\vec{k}.\vec{r}}$$

As will become apparent, we need to count how thickly these states are distributed, both in momentum space (or *k*-space) and in energy. We'll begin by reviewing the one-dimensional problem the three-dimensional case is a simple generalization.

Recall that for particles in *one* dimension confined to a line of length *L* with periodic boundary conditions, the allowed values of wave number *k* were given by $e^{ikL} = 1$, so $k = 2n\pi / L$ with *n* an integer. Thus considering only intervals $\Delta k >> 2\pi / L$, the "density of states" in *k* is L/2π: an interval of length Δk contains $(L/2\pi)\Delta k = \rho(k)\Delta k$ states, where here $\rho(k) = L/2\pi$. The density of states in *energy*, $\rho(E)$, follows from differentiating $E = \hbar^2 k^2 / 2m$. Writing $\Delta E = (\hbar^2 k / m)\Delta k$, gives the incremental change ΔE in *E* for a given incremental change Δk in *k*, so the two intervals ΔE and Δk must contain the same number of states, that is, $\rho(E)\Delta E = \rho(k)\Delta k$. It then follows from $\rho(k) = L/2\pi$ that the one-dimensional density of states in energy

$$\rho_{1D}(E) = (L/2\pi)(m/\hbar^2 k) = (L/2\pi\hbar)\sqrt{m/2E}$$

Note this *one-dimensional* density of states goes to infinity as *E* goes to zero.

In *three* dimensions, with a cube of side *L* and periodic boundary conditions, the density of states in *k*-space is $(L/2\pi)^3$. The allowed states can be visualized as the points of a cubic lattice, $(k_x, k_y, k_z) = \frac{2\pi}{L}(n_x, n_y, n_z)$, the *n*'s being integers, so each allowed state has associated with it the volume of a small cube $(2\pi/L)^3$.

To find the three-dimensional density of states in energy, using $E = \hbar^2 \vec{k}^2 / 2m$, again $\Delta E = \left(\hbar^2 k / m\right) \Delta k$ but now to find the number of states in a small energy range we must multiply by $4\pi k^2$, since the states in the energy range lie between two close concentric spheres in k-space. This gives

$$\rho(E) = (L/2\pi)^3 4\pi k^2 \left(m/\hbar^2 k\right) = (L/2\pi)^3 4\pi k \left(m/\hbar^2\right) = \left(V/2\pi^2\right)\left(m/\hbar^3\right)\sqrt{2mE}$$

Notice that in contrast to the one-dimensional case, the three-dimensional density of states goes to *zero* at zero energy. (*Exercise*: What happens in two dimensions?)

(Of course, if we are detecting the ejected electron with apparatus restricted to a solid angle $d\Omega$, the 4π is replaced by $d\Omega$.)

The orthogonality condition between the plane wave states is

$$\left\langle \vec{k} \middle| \vec{k}' \right\rangle = \delta_{\vec{k},\vec{k}'}$$

the ordinary Kronecker delta function not Dirac's since the k's are an enumerated set,

$$\left(k_x, d_y, k_z\right) = \frac{2\pi}{L}\left(n_x, n_y, n_z\right) \text{ the } n\text{'s being integers.}$$

FINDING THE MATRIX ELEMENT

The ground state wave function for hydrogen is

$$|100\rangle = \sqrt{\frac{1}{\pi a_0^3}} e^{-r/a_0}.$$

The matrix element entering Fermi's Golden Rule is therefore:

$$R_{i\to f} = \frac{2\pi}{\hbar}\left|\langle f|H^1|i\rangle\right|^2 \delta\left(E_f - E_i - \hbar\omega\right)$$

$$\langle \vec{k}_f | \left(\frac{e}{2mc}\right) e^{i\vec{k}\cdot\vec{r}} \vec{A}_0 \cdot \vec{p} |100\rangle = \int d^3r (1/L)^{3/2} e^{-i\vec{k}_f\cdot\vec{r}} \left(\frac{e}{2mc}\right) e^{i\vec{k}\cdot\vec{r}} \vec{A}_0 \cdot \left(-i\hbar\vec{\nabla}\right) \sqrt{\frac{1}{\pi a_0^3}} e^{-r/a_0}$$

Actually the $e^{i\vec{k}\cdot\vec{r}}$ term is not very important the wavelength of incoming photons for the usual photoelectric effect is far greater than the size of the hydrogen atom in its ground state (which our integral is limited to) so $e^{i\vec{k}\cdot\vec{r}} \cong 1$, and we can drop that term.

One point we've overlooked is that the electromagnetic wave has a *magnetic* field just as strong as the electric field, so what about the interaction of this magnetic field with the electron's magnetic moment? This turns out to be much weaker than the $\left(\frac{e}{2mc}\right)\vec{A}_0 \cdot \vec{p}$ term: the magnetic interaction

$$\vec{\mu}_B \cdot \vec{B} = \left(\frac{e}{2mc}\right)\vec{S} \cdot \vec{B},$$

and the ratio of this magnetic contribution to the electric one is

$$\frac{\left(\frac{e}{2mc}\right)\vec{S} \cdot \vec{B}}{\left(\frac{e}{2mc}\right)\vec{A}_0 \cdot \vec{p}} = \frac{\hbar \cdot \vec{\sigma} \times \vec{A}}{\vec{A} \cdot \vec{p}} = \frac{\hbar k}{p},$$

with $p \sim \hbar / a_0$, so this ratio is of order a_0 / λ, l being the wavelength of the incoming light, around 100 nm to ionize hydrogen. So, we can safely ignore the magnetic interaction.

This interaction Hamiltonian $H^1 = \left(\frac{e}{2mc}\right)\vec{A}_0 \cdot \vec{p}e^{-i\omega t}$ is called the *dipole approximation,* because it can also be written in terms of the atom's dipole moment as follows:

If $|i,\rangle | f\rangle$ are eigenstates of $H = \vec{p}^2 / 2m + V(\vec{r})$, then from $[\vec{r}, \vec{p}] = i\hbar, [\vec{r}, H] = (i\hbar / m)\vec{p}$,

and

$$\langle f|\vec{p}|i\rangle = (m/i\hbar)\langle f|\vec{r}H - H\vec{r}|i\rangle$$
$$= (m/i\hbar)(E_i - E_f)\langle f|\vec{r}|i\rangle$$
$$= im\omega\langle f|\vec{r}|i\rangle.$$

This means that

$$\langle f|H^1(t)|i\rangle = \left(\frac{e}{2mc}\right)\vec{A}_0 e^{-i\omega t}\cdot\langle f|\vec{p}|i\rangle = \left(\frac{e}{2mc}\right) im\omega\vec{A}_0 e^{-i\omega t}\cdot\langle f|\vec{r}|i\rangle,$$

then recall $\vec{E} = -(1/c)\partial\vec{A}/\partial t = (i\omega/2c)\vec{A}_0 e^{-i\omega t}$, from which $\langle f|H^1(t)|i\rangle = \langle f|-\vec{\mu}\cdot\vec{E}(t)|i\rangle$, with $\vec{\mu} = -e\vec{r}$, the atom's electric dipole moment.

Nevertheless, for the particular interaction we are considering here, the $\vec{p} = -i\hbar\vec{\nabla}$ representation proves more convenient. (We'll use the representation in later work.)

We must evaluate: $(1/L)^{3/2}\left(\frac{e}{2mc}\right)\sqrt{\frac{1}{\pi a_0^3}}\int d^3r e^{-i\vec{k}_f\cdot\vec{r}}\vec{A}_0\cdot\left(-i\hbar\vec{\nabla}\right)e^{-r/a_0}$.

Integration by parts gives the gradient operator acting on the plane wave state,

$$\int d^3r e^{-i\vec{k}_f\cdot\vec{r}}\vec{A}_0\cdot\left(-i\hbar\vec{\nabla}\right)e^{-r/a_0} = -\left(\vec{A}_0\cdot\vec{p}_f\right)\int d^3r e^{-i\vec{k}_f\cdot\vec{r}}e^{-r/a_0}.$$

The integral is now a Fourier transform of the hydrogen ground state wave function, and is straightforward: choose the z-axis in the direction of $\vec{k}_f$, the φ-integration gives 2π, the θ-integration has $\sin\theta d\theta = -d(\cos\theta)$, etc. The result is

$$(8\pi/a_0)/\left(a_0^{-2}+k_f^2\right)^2$$

Finally, we can put this into Fermi's Golden Rule:

$$R_{i\to f} = \frac{2\pi}{\hbar}\left|(1/L)^{3/2}\left(\frac{e}{2mc}\right)\sqrt{\frac{1}{\pi a_0^3}}\left(\vec{A}_0\cdot\vec{p}_f\right)\left(\frac{8\pi/a_0}{\left(a_0^{-2}+k_f^2\right)^2}\right)\right|^2\delta(E_f - E_i - \hbar\omega).$$

To detect the ejected electron, we will have a detector sensitive to some small solid angle, $d\Omega$, not to some precise value of $\vec{p}_f$. There will also be some tiny uncertainty in $\left|\vec{p}_f\right|$, equivalent to an energy uncertainty, because for one thing the ejection takes place after a finite time. This means the δ-function actually has finite width, and by taking our normalizing box big enough, there will be many states within this width so, effectively, the δ-function is measuring the density of possible outgoing states. Recall the density of states in energy for outgoing solid angle $d\Omega$ is

$$\rho(E, d\Omega) = (L/2\pi)^3 k^2 (m/\hbar^2 k) d\Omega = (L/2\pi)^3 k (m/\hbar^2) d\Omega,$$

giving

$$R_{i\to f} = \frac{2\pi}{\hbar}\left|(1/L)^{3/2}\left(\frac{e}{2mc}\right)\sqrt{\frac{1}{\pi a_0^3}}\left(\vec{A}_0\cdot\vec{p}_f\right)\left(\frac{8\pi/a_0}{\left(a_0^{-2}+k_f^2\right)^2}\right)\right|^2 (L/2\pi)^3 k_f (m/\hbar^2) d\Omega.$$

Notice first that the L^3 terms cancel, reassuringly, our result cannot depend on the size of the box chosen for the plane wave states. Writing $p_f = \hbar k_f$, and of course $p_f^2/2m = E_i + \hbar\omega$, we find

$$R_{i\to f} = \frac{4mp_f}{\pi a_0 \hbar^4}\left(\frac{e}{mc}\right)^2 \left(\vec{A}_0\cdot\vec{p}_f\right)^2 \left(\frac{1}{a_0^{-2}+\left(p_f/\hbar\right)^2}\right)^4 d\Omega.$$

Note that the rate is angle-dependent, since $\left(\vec{A}_0\cdot\vec{p}_f\right)^2 = A_0^2 p_f^2 \cos^2\theta$: ejection is most likely parallel to the electric field. The *total* ionization rate is given by integrating the rate over all angles, and on the unit sphere $\overline{\cos^2\theta} = \overline{z^2} = 1/3$, so in the above,.

THE PHOTOELECTRIC CROSS SECTION

Imagine now sending this radiation into a gas of hydrogen atoms, many of them, but not enough to shade each other from the radiation significantly. Energy will be absorbed from the beam as atoms ionize. What is the rate at which the beam loses energy? A convenient way of visualizing this rate of loss of energy is to replace each atom by a tiny perfectly absorbent disc oriented with its normal parallel to the beam, the size of these discs such that the beam loses energy at the same rate as it would by ionization. The area of the disc equivalent to one atom is called the *photoelectric cross section*.

The energy density in the beam of radiation is

$$\frac{1}{8\pi}\left(\left|\vec{E}\right|^2+\left|\vec{B}\right|^2\right)=\frac{1}{8\pi}\left(2\frac{\omega^2}{c^2}\vec{A}_0^2\cos^2\left(\vec{k}\cdot\vec{r}-\omega t\right)\right)$$

Denoting the photoelectric cross section by,

σ, energy absorbed per second = σ × c × energy density

and averaging $\cos^2$, this gives the energy absorption rate per atom to be $A_0^2\omega^2\sigma/8\pi c$. However, if the rate of ionization of one atom is $R_{i\to f}$, and that ionization takes energy from the beam $\hbar\omega$, the rate of energy absorption is just $\hbar\omega R_{i\to f}$, so the ionization cross section is given by

$A_0^2\omega^2\sigma/8\pi c=\hbar\omega R_{i\to f}$ This gives

$$\sigma=\hbar\omega R_{i\to f}$$

$$=\frac{8\pi c}{A_0^2\omega^2}\hbar\omega\frac{4mp_f}{\pi a_0\hbar^4}\left(\frac{e}{mc}\right)^2\frac{4\pi A_0^2p_f^2}{3}\left(\frac{1}{a_0^{-2}+\left(p_f/\hbar\right)^2}\right)^4$$

$$=\frac{128}{\omega}\frac{e^2}{a_0\hbar^3}\frac{\pi p_f^3}{3mc}\left(\frac{1}{a_0^{-2}+\left(p_f/\hbar\right)^2}\right)^4$$

APPENDIX: THE GOLDEN RULE DELTA FUNCTION AND THE DENSITY OF STATES

For our big box model, the states are infinite in number, but can be counted by going outwards from the origin in k-space, and adopting some convention for ordering those of equal energy. We can label the states with $\vec{n}=\left(n_x,n_y,n_z\right)$, a vector with integer components placing the state in k-space, and denote its energy $E_{\vec{n}}$. The contribution of this state to the density of states is a Dirac δ-function $\delta\left(E-E_{\vec{n}}\right)$, that is to say, this state contributes 1 to the density of states at the point $E_{\vec{n}}$ on the energy axis. Therefore the density of states in energy $\rho(E)=\sum_{\vec{n}}\delta\left(E-E_{\hbar}\right)$, well approximated by the smooth function we derived above.

Now consider the integral over final plane wave states needed in the evaluation of the Golden Rule formula. *That* δ-function has finite width (from the time-energy uncertainty principle), so by taking our big box big enough we can have many plane wave states within the width of the Golden Rule δ-function: to picture this, let's represent it by a function equal to Δ over an interval, zero otherwise. Then integrating this Golden Rule δ-function with $\rho(E)$ will give a contribution Δ from each state inside the interval of width $1/\Delta$. If the states were uniformly distributed in energy, this would give the total number of states in an interval of unit energy and that is the definition of the density of states.

Chapter 11

Quantizing Radiation

INTRODUCTION

In analyzing the photoelectric effect in hydrogen, we derived the rate of ionization of a hydrogen atom in a monochromatic electromagnetic wave of given strength, and the result we derived is in good agreement with experiment. Recall that the interaction Hamiltonian was

$$H^1 = \left(\frac{e}{mc}\right)\cos\left(\vec{k}\cdot\vec{r} - \omega t\right)\vec{A}_0\cdot\vec{p}$$

$$= \left(\frac{e}{2mc}\right)\left(e^{i(\vec{k}\cdot\vec{r}-\omega t)} + e^{-i(\vec{k}\cdot\vec{r}-\omega t)}\right)\vec{A}_0\cdot\vec{p}.$$

and we dropped the $e^{i\omega t}$ term because it would correspond to the atom giving energy to the field, and our atom was already in its ground state. However, if we go through the same calculation for an atom *not* initially in the ground state, then indeed an electromagnetic wave of appropriate frequency will cause a transition rate to a lower energy state, and $e^{i\omega t}$ is the relevant term.

But this is not the whole story. An atom in an excited state will eventually emit a photon and go to a lower energy state, even if there is *zero* external field. Our analysis so far does not predict this obviously, the interaction written above is only nonzero if $\vec{A}$ is nonzero! So what are we missing?

Essentially, the answer is that the electromagnetic field itself is quantized. Of course, we know that, it's made up of photons. Recall Planck's successful analysis of radiation in a box: he considered all possible normal modes for the radiation, and asserted that a mode of energy *w* could only gain or lose energy in amounts $\hbar\omega$. This led to the correct formula for black body radiation, then Einstein proved that the same assumption, with the same $\hbar$, accounted for the photoelectric effect. We now understand that these modes of oscillation of radiation are just simple harmonic oscillators, with energy $\left(n+\frac{1}{2}\right)\hbar\omega$, and, just as a mass on a spring oscillator has fluctuations in the ground state $\langle x\rangle = 0$, but $\left\langle x^2\right\rangle \neq 0$, for these electromagnetic modes $\langle A\rangle = 0$ but $\left\langle \vec{A}^2\right\rangle \neq 0$.

These fluctuations in $\vec{A}$ mean the interaction Hamiltonian is momentarily nonzero, and therefore can cause a transition.Therefore, to find the spontaneous transition rate (as it's called) for an atom in a zero (classically speaking) electromagnetic field, we need to express the electromagnetic field in terms of normal modes (we'll take a big box), then quantize these modes as quantum simple harmonic oscillators, introducing raising and lowering operators for each oscillator (these will be photon creation and annihilation operators) then construct the appropriate quantum operator expression for $\vec{A}$ to put in the electron-radiation interaction Hamiltonian.

The bras and kets will now be quantum states of the electron *and* the radiation field, in contrast to our analysis of the classical field above, where the radiation field didn't change. (Of course, it did, really, in that it lost one photon, but in the classical limit there are infinitely many photons in each mode, so that wouldn't register.)

We use the Coulomb gauge $\vec{\nabla}\cdot\vec{A} = 0$, and of course $\vec{A}$ satisfies $\nabla^2\vec{A} - \frac{1}{c^2}\frac{\partial^2\vec{A}}{\partial t^2} = 0$.

Taking for convenience periodic boundary conditions in the big box, we can write $\vec{A}$ (classically) as a Fourier series at $t = 0$:

$$\vec{A}(\vec{r},t=0)=\frac{1}{\sqrt{V}}\sum_{\vec{k}}\sum_{\alpha=1,2}\left(c_{\vec{k},\alpha}(0)\vec{\varepsilon}_\alpha e^{i\vec{k}\cdot\vec{r}}+c^*_{\vec{k},\alpha}(0)\vec{\varepsilon}^*_\alpha e^{-i\vec{k}\cdot\vec{r}}\right)$$

The time-dependence is given by putting in the whole plane wave: $e^{i\bar{k}\bar{r}} \rightarrow e^{i(\bar{k}\bar{r}-\omega t)}$, which time dependence can be taken into the coefficient, $c_{\bar{k},\alpha}(t)=c_{\bar{k},\alpha}(0)e^{-i\omega t}$, so

$$\vec{A}(\vec{r},t)=\frac{1}{\sqrt{V}}\sum_{\vec{k}}\sum_{\alpha=1,2}\left(c_{\vec{k},\alpha}(t)\vec{\varepsilon}_\alpha e^{i\vec{k}\cdot\vec{r}}+c^*_{\vec{k},\alpha}(t)\vec{\varepsilon}^*_\alpha e^{-i\vec{k}\cdot\vec{r}}\right)$$

The vector $\vec{\varepsilon}_\alpha$ is the polarization of the plane wave. It's in the same direction as the electric field. Actually it varies with $\vec{k}$, because from $\vec{\nabla}\cdot\vec{A}=0$, it's perpendicular to $\vec{k}$. That is, for a given $\vec{k}$ there are two independent polarizations. For $\vec{k}$ along the z-axis, they could be along the *x*- and *y*-axes, these would be called linear polarization, and is the standard approach. But we could also take the vectors $\left(1/\sqrt{2}\right)(1,\pm i,0)$. These correspond to *circular polarization*: equal *x*-and *y*-components but with the *y*-component 90 degrees ahead in phase. You may recognize the vectors $\left(1/\sqrt{2}\right)(1,\pm i,0)$ as the eigenvectors for the rotation operator around the z-axis the circularly polarized beam carries angular momentum, $\pm\hbar$ per photon, pointed along the direction of motion.

The energy density $\frac{1}{8\pi}\left(\left|\vec{E}\right|^2+\left|\vec{B}\right|^2\right)$ can be expressed as a sum over the individual $\left(\vec{k},\vec{\varepsilon}\right)$ modes.

Writing the electric and magnetic fields in terms of the vector potential,

$\vec{E}=-(1/c)\partial\vec{A}/\partial t,\ \vec{B}=\vec{\nabla}\times\vec{A}.$

where

$$\vec{A}(\vec{r},t)=\frac{1}{\sqrt{V}}\sum_{\vec{k}}\sum_{\alpha=1,2}\left(c_{\vec{k},\alpha}(t)\vec{\varepsilon}_\alpha e^{i\vec{k}\cdot\vec{r}}+c^*_{\vec{k},\alpha}(t)\vec{\varepsilon}^*_\alpha e^{-i\vec{k}\cdot\vec{r}}\right)$$

and thereby expressing the total energy $\frac{V}{8\pi}\left(\overline{|\vec{E}|^2+|\vec{B}|^2}\right)=\frac{V}{4\pi}\left(\frac{\omega}{c}\right)^2\overline{|\vec{A}|^2}$ in terms of the $(\vec{k},\vec{\varepsilon})$ amplitudes $c^{\bullet}_{\bar{k},\alpha}(t), c_{\bar{k},\alpha}(t)$, then integrating the energy density over the whole large box the cross terms disappear from the orthogonality of the different modes and the total energy in the box the Hamiltonian is:

$$H=\frac{1}{2\pi}\sum_{\bar{k}}\sum_{\alpha}\left(\frac{\omega}{c}\right)^2 c^{\bullet}_{\bar{k},\alpha}c_{\bar{k},\alpha}.$$

Note that although the Hamiltonian is (of course) time independent, the *coefficients* $C_{\bar{k},\alpha}$ here *are* time dependent,.

$$c_{\bar{k},\alpha}(t)=c_{\bar{k},\alpha}(0)e^{-i\omega t}$$

But this is formally *identical* to a set of simple harmonic oscillators! Recall that for the classical oscillator, $p^2+(m\omega x)^2=2mE$, the vector $z=m\omega x+ip$ has time dependence $z(t)=z_0e^{-i\omega t}$, and the oscillator energy is proportional to $z\cdot z$ (*x, p* are the usual conjugate variables). Clearly, $C_{\bar{k},\alpha}(t)$ here corresponds to *z*(*t*): same time dependence, same Hamiltonian. Therefore the real and imaginary parts of $C_{\bar{k},\alpha}(t)$ must *also* be conjugate variables, which can therefore be quantized exactly as for the simple harmonic oscillator.

From

$$\vec{A}(\vec{r},t)=\frac{1}{\sqrt{V}}\sum_{\bar{k}}\sum_{\alpha=1,2}\left(c_{\bar{k},\alpha}(t)\,\vec{\varepsilon}_{\alpha}e^{i\bar{k}\vec{r}}+c^{\bullet}_{\bar{k},\alpha}(t)\,\vec{\varepsilon}^{\,\bullet}_{\alpha}e^{-i\bar{k}\vec{r}}\right)$$

we see that the real part of $C_{\bar{k},\alpha}(t)$ basically gives the contribution of the $\vec{k},\alpha$ oscillator to $\vec{A}(\vec{r},t)$, and, recalling the time dependence $C_{\bar{k},\alpha}(t)=C_{\bar{k},\alpha}(0)e^{-i\omega t}$, the imaginary part is proportional to the contribution to $\partial\vec{A}(\vec{r},t)/\partial t$, that is $\vec{E}(\vec{r},t)$, to. Essentially, then, the real part of $C_{\bar{k},\alpha}(t)$, proportional to the $\vec{k},\alpha$ Fourier component of the vector potential $\vec{A}$, is what corresponds to displacement x in a 1-D simple harmonic oscillator, and the imaginary part of $C_{\vec{k},\alpha}(t)$, the $\vec{k},\alpha$ Fourier component of $\vec{E}$, corresponds to the momentum in the simple harmonic oscillator.

To carry out the quantization, we must express the classical Hamiltonian

$$H = \frac{1}{2\pi}\sum_{\vec{k}}\sum_{\alpha}\left(\frac{\omega}{c}\right)^2 c^{*}_{\vec{k},\alpha}c_{\vec{k},\alpha}$$

in the form

$$H = \sum_{\vec{k}}\sum_{\alpha}\tfrac{1}{2}\left(P^2_{\vec{k},\alpha} + \omega^2 Q^2_{\vec{k},\alpha}\right)$$

with $P_{\vec{k},\alpha}, Q_{\vec{k},\alpha}$ being the imaginary and real parts of the oscillator amplitude $C_{\vec{k},\alpha}(t)$ (scaled appropriately) exactly parallel to the standard treatment of the simple harmonic oscillator:

$$Q_{\vec{k},\alpha} = \frac{1}{c\sqrt{4\pi}}\left(c_{\vec{k},\alpha} + c^{*}_{\vec{k},\alpha}\right), \quad P_{\vec{k},\alpha} = -\frac{i\omega}{c\sqrt{4\pi}}\left(c_{\vec{k},\alpha} - c^{*}_{\vec{k},\alpha}\right).$$

From the time-dependence $C_{\vec{k},\alpha}(t) = C_{\vec{k},\alpha}(0)e^{-i\omega t}$, these (classical) variables P, Q are *canonical*:

$$\frac{\partial H}{\partial Q_{\vec{k},\alpha}} = -\dot{P}_{\vec{k},\alpha}, \quad \frac{\partial H}{\partial P_{\vec{k},\alpha}} = \dot{Q}_{\vec{k},\alpha}$$

The Hamiltonian can now be quantized by the standard procedure. The pairs of canonical variables P, Q (one pair to each mode $\vec{k},\alpha$) become operators, the Poisson brackets become commutators, the scale determined by Planck's constant:

$$\left[Q_{\vec{k},\alpha}, P_{\vec{k}',\alpha'}\right] = i\hbar\delta_{\vec{k},\vec{k}'}\delta_{\alpha,\alpha'}.$$

$$a_{\vec{k},\alpha} = \frac{1}{\sqrt{2\hbar\omega}}\left(\omega Q_{\vec{k},\alpha} + iP_{\vec{k},\alpha}\right)$$

$$a^\dagger_{\vec{k},\alpha} = \frac{1}{\sqrt{2\hbar\omega}}\left(\omega Q_{\vec{k},\alpha} - iP_{\vec{k},\alpha}\right)$$

These satisfy $\left[a, a^\dagger\right] = 1$.

(Notice that the annihilation operator $a_{\vec{k},\alpha}$ is nothing but the operator representation of the classical complex amplitude $c_{\vec{k},\alpha}$, with an extra factor to make it dimensionless,

$$c_{\vec{k},\alpha} \to c\sqrt{\frac{2\pi\hbar}{\omega}}\, a_{\vec{k},\alpha}.$$

We discussed this same equivalence in the lecture on coherent states, which were eigenstates of the annihilation operator.)

Following the standard simple harmonic oscillator development, the operator $\hat{n}_{\vec{k},\alpha} = \alpha^\dagger_{\vec{k},\alpha}\alpha_{\vec{k},\alpha}$ has eigenstates with integer eigenvalues, $\hat{n}|n\rangle = n|n\rangle$, the contribution to the Hamiltonian from the mode $\vec{k},\alpha$ is just $H_{\vec{k},\alpha} = \left(\hat{n}_{\vec{k},\alpha} + \frac{1}{2}\right)\hbar\omega$, and $a^\dagger|n\rangle = \sqrt{n+1}\,|n+1\rangle$, $a|n\rangle = \sqrt{n}\,|n-1\rangle$.

The bottom line is: the classical plane wave expansion of $\vec{A}$, with wave amplitudes $C_{\vec{k},\alpha}(t)$

$$\vec{A}(\vec{r},t) = \frac{1}{\sqrt{V}}\sum_{\vec{k}}\sum_{\alpha=1,2}\left(c_{\vec{k},\alpha}(t)\,\vec{\varepsilon}_\alpha e^{i\vec{k}\cdot\vec{r}} + c^*_{\vec{k},\alpha}(t)\,\vec{\varepsilon}^{\,*}_\alpha e^{-i\vec{k}\cdot\vec{r}}\right)$$

is replaced on quantization by a parallel *operator* expansion, the wave amplitude $C_{\vec{k},\alpha}(t)$ becoming the (scaled) annihilation operator:

$$\vec{A}(\vec{r},t) = \frac{1}{\sqrt{V}}\sum_{\vec{k}}\sum_{\alpha=1,2} c\sqrt{\frac{2\pi\hbar}{\omega}}\left(a_{\vec{k},\alpha}(t)\,\vec{\varepsilon}_\alpha e^{i\vec{k}\cdot\vec{r}} + a^\dagger_{\vec{k},\alpha}(t)\,\vec{\varepsilon}^{\,*}_\alpha e^{-i\vec{k}\cdot\vec{r}}\right).$$

SPONTANEOUS EMISSION

However, this exact correspondence with the classical result does *not* hold for photon emission! In that case, the atom *adds* a photon to a mode which already contains n photons, say, and the relevant matrix element is $a^\dagger|n\rangle = \sqrt{n+1}|n+1\rangle$, so the equivalent classical vector $\vec{A}_0$ is $c\sqrt{\frac{(n_{\vec{k},\alpha}+1)2\pi\hbar}{\omega V}}\vec{\varepsilon}_\alpha$. This is nonzero even if $n_{\vec{k},\alpha}$ is zero hence *spontaneous* emission.

For spontaneous emission, then, the relevant matrix element is

$$\langle 100;1|\left(\frac{e}{mc}\right)e^{-i\vec{k}\cdot\vec{r}}c\sqrt{\frac{2\pi\hbar}{\omega}}a^\dagger_{\vec{k},\alpha}\frac{\vec{\varepsilon}\cdot\vec{p}}{\sqrt{V}}|21m;0\rangle.$$

The density of outgoing states for the emitted photon, taking box normalization with periodic boundary conditions as usual, is

$$\frac{V}{(2\pi)^3}k^2dkd\Omega = \frac{V}{(2\pi)^3}\frac{\omega^2 d\omega d\Omega}{c^3} = \frac{V}{(2\pi)^3}\frac{\omega^2 dEd\Omega}{\hbar c^3}$$

so the density of states in energy contribution to the Golden Rule delta function is $\frac{V}{(2\pi)^3}\frac{\omega^2 d\Omega}{\hbar c^3}$, and the photon emission rate with polarization $\vec{\varepsilon}$ into a solid angle $d\Omega$ will be:

$$\frac{2\pi}{\hbar}\left|\langle 100;1|\left(\frac{e}{mc}\right)e^{-i\vec{k}\cdot\vec{r}}c\sqrt{\frac{2\pi\hbar}{\omega}}a^\dagger_{\vec{k},\alpha}\frac{\vec{\varepsilon}\cdot\vec{p}}{\sqrt{V}}|21m;0\rangle\right|^2\frac{V}{(2\pi)^3}\frac{\omega^2 d\Omega}{\hbar c^3}.$$

One slight difference in evaluating the matrix element from our treatment of the photoelectric effect is in the representation of the dipole interaction. Recall that there we gave the equivalent forms

$$\langle f|H^1|i\rangle = \left(\frac{e}{mc}\right)\vec{A}_0\cdot\langle f|\vec{p}|i\rangle e^{-i\omega t} = \left(\frac{e}{mc}\right)im\omega\vec{A}_0\cdot\langle f|\vec{r}|i\rangle e^{-i\omega t}$$

and used the $\vec{p}$ representation because the outgoing photoelectron was taken to be in a plane wave state, an eigenstates of $\vec{p}$. But for spontaneous emission, the electron goes from one bound state to another, so the $\vec{r}$ form gives a more immediate picture of the interacting dipole with the external field, and in fact the integration between the states is generally a little more direct.

So in the matrix element we make the substitution $\vec{\varepsilon} \cdot \vec{p} \rightarrow imm\omega\vec{\varepsilon} \cdot \vec{r}$, and must then evaluate the atomic matrix element $\langle 100|\vec{\varepsilon} \cdot \vec{r}|21m\rangle$. The natural way to do this is to express the vectors in terms of spherical harmonics, that is, to write them as spherical vectors,

$$r_1^{\pm 1} = \mp(x \pm iy)/\sqrt{2} = r\sqrt{4\pi/3}Y_1^{\pm 1},\ r_1^0 = z = r\sqrt{4\pi/3}Y_1^0$$

and similarly for $\vec{\varepsilon}$ The integrals are then straightforward but tedious An amusing point made by Sakurai is that the total transition probability for spontaneous emission is $\frac{1}{137}\frac{4}{3}\frac{\omega^3}{c^2}|\langle 100|\vec{x}|21m\rangle|^2$ and this same expression was obtained using the Correspondence Principle by Heisenberg, before quantum field theory was invented.

The calculated lifetime of the $n = 2$ state is 1.6×10^{-9} seconds.

Chapter 12

Scattering Theory

INTRODUCTION

Almost everything we know about nuclei and elementary particles has been discovered in scattering experiments, from Rutherford's surprise at finding that atoms have their mass and positive charge concentrated in almost point-like nuclei, to the more recent discoveries, on a far smaller length scale, that protons and neutrons are themselves made up of apparently point-like quarks.

The simplest model of a scattering experiment is given by solving Schrödinger's equation for a plane wave impinging on a localized potential. A potential $V(r)$ might represent what a fast electron encounters on striking an atom, or an alpha particle a nucleus. Obviously, representing any such system by a potential *is* only a beginning, but in certain energy ranges it is quite reasonable, and we have to start somewhere!

The basic scenario is to shoot in a stream of particles, all at the same energy, and detect how many are deflected into a battery of detectors which measure angles of deflection. We assume all the ingoing particles are represented by wavepackets of the same shape and size, so we should solve Schrödinger's time-dependent equation for such a wave packet and find the probability amplitudes for outgoing waves in different directions at some later time after scattering has taken place. But we adopt a simpler approach: we assume the wavepacket has a well-defined energy (and hence momentum), so it is *many* wavelengths long. This means that during the scattering process it looks a lot like a plane wave, and for a period of time the scattering is time independent. We assume, then, that the problem is well approximated by solving the time-*independent* Schrödinger equation with an ingoing plane wave. This is much easier!

All we can detect are outgoing waves far outside the region of scattering. For an ingoing plane wave $e^{i\mathbf{k}.\mathbf{r}}$, the wavefunction *far away from the scattering region* must have the form

$$\psi_k(r) = e^{ikr} + f(\theta,\varphi)\frac{e^{ik\gamma}}{r}$$

where q, φ are measured with respect to the ingoing direction.

Note that the *scattering amplitude* $f(q,\varphi)$ has the dimensions of length.

We don't worry about overall normalization, because what is relevant is the *fraction* of the incoming beam scattered in a particular direction, or, to be more precise, into a small solid angle $d\Omega$ in the direction q, φ. The ingoing particle current (with the above normalization) is $\hbar j/m = v$ through unit area perpendicular to the ingoing beam, the outgoing current into the small angle dΩ is $(\hbar k/m)|f(\theta,\varphi)|^2\, d\Omega$. It is evident that this outgoing current corresponds to the original ingoing current flowing through a perpendicular area of size, and

$$\frac{d\sigma}{d\Omega}|f(\theta,\varphi)|^2$$

is called the *differential cross* **section** for scattering in the direction (q, φ).

THE TIME-INDEPENDENT DESCRIPTION

We shall review the time-independent formulation of scattering theory, first as it is presented in Baym, in terms of the standard Schrödinger equation wavefunctions, then do the same thing a la Sakurai, in the more formal, but of course equivalent, language of bras and kets. The Schrödinger wavefunction approach is an easier introduction, but the formal language is more convenient for analyzing the structure of higher order terms.

Actually, Baym's treatment isn't *quite* time-independent, in that he uses an ingoing wavepacket, but it is one of great length, well approximated by a plane wave. Sakurai goes straight to the plane wave, and we do too. This case is very reminiscent of one-dimensional scattering, in which a plane wave from the left generates outgoing waves in both directions, and the amplitudes can be calculated from the Schrödinger equation for a single energy eigenstate. The only difference is that in 3D there will be outgoing waves in all directions.

Following Baym, Schrödinger's equation is:

$$\left(\frac{\hbar^2}{2m}\nabla^2 + E_k\right)\psi_{\mathbf{k}}(\mathbf{r}) = V(\mathbf{r})\psi_{\mathbf{k}}(\mathbf{r}), \quad \text{where } E_k = \frac{\hbar^2 k^2}{2m}.$$

This ψ_{k} we take to have an incoming plane wave component $e^{i\mathrm{k.r}}$. Overall normalization is irrelevant, since the differential cross-section depends only on the *ratio* of the scattered wave amplitude to that of the ingoing wave.

The standard approach to an equation like the one above is to transform it into an integral equation using Green's functions. If $V(\mathrm{r})$ is small (just how small it has to be will become clear later) the integral equation can then be solved by iteration.

The Green's function $G(\mathrm{r}, \mathrm{k})$ is essentially the inverse of the differential operator,

$$\left(\frac{\hbar^2}{2m}\nabla^2 + E_k\right)G(\mathbf{r},\mathbf{k}) = \delta(\mathbf{r}).$$

This is not a mathematically unique definition: clearly, we can add to $G(\mathrm{r}, \mathrm{k})$ any solution of the homogeneous equation

$$\left(\frac{\hbar^2}{2m}\nabla^2 + E_k\right)\varphi(\mathbf{r},\mathbf{k}) = 0,$$

for example, the incoming plane wave.

If we write the integral equation

$$\psi_k(r) = e^{ikr} + \int d^3r' G(r-r') V(r') \psi_k(r')$$

this $\psi_k(r)$ is certainly a solution to the original Schrödinger equation, as is easily checked by applying the operator

$$\left(\frac{\hbar^2}{2m}\nabla^2 + E_k\right)$$

to both sides of the equation.

The integral equation can be formally solved by iteration, and for "small" *V* the solution will converge. But this won't really do remember, we haven't a unique *G*(r, k)! We have to fix *G*(r, k) by connecting better with the scattering problem we're trying to solve.

We know our solution has a *single* ingoing plane wave, and outgoing waves in all other directions, generated by the interaction of the plane wave with the potential. But the Schrödinger equation could equally describe *ingoing* waves in the other directions. In defining the Green's function and writing the integral equation, we have nowhere specified the distant form of the wavefunction, that is, we have not required that the Green's function on the right hand side of the integral equation only generate *outgoing* waves. To see how to do this, we must write the Green's function itself as a sum over waves, in other words a Fourier transform, and see how to eliminate the unphysical (for the present problem) incoming waves in that sum.

The explicit form of the Green's function is

$$G(r,k) = \int \frac{d^3k'}{(2\pi)^3} \frac{e^{ik'r}}{E_k - \frac{\hbar^2 k'^2}{2m}} = -\frac{m}{2\pi^2 i r \hbar^2} \int_{-\infty}^{\infty} \frac{k'dk' e^{ik'r}}{k'^2 - k^2}.$$

Note that *G*(*r*, *k*) only depends on k through E_k, and only on r through *r*, since the integration over k′ is over all directions. It is easy to verify that this Green's function satisfies the differential equation, by applying the differential operator to the first integral above: the result is to cancel the denominator in the integral, leaving just $\int \frac{d^3k'}{(2\pi)^3} e^{i\mathbf{k}'\mathbf{r}}$, which is the d-function in r.

To get the second form of $G(r, k)$ in the equation above, we first do the angular integration $d(\cos\theta)$ to get $\left(e^{ik'r} - e^{-ikr}\right)/iK'r$, then rearrange the integral over the $-e^{-ik'r}$ term by switching the sign of k', so it becomes an integral from $-\infty$ to 0 instead of 0 to ∞. Then we add the two terms (the $e^{ik2\,r}$ and the $-e^{-ik2\,r}$) together to give an integral from $-\infty$ to ∞. This integral from $-\infty$ to ∞ is then done by contour integration at least, after we've figured out what to do about the singularities at $k' = \pm k$.

For the integral to be defined, the contour must be distorted slightly so it bypasses these poles.

It is at this point we feed in our physical knowledge of the situation: that in the scattering process, the second term in

$$\psi_{\mathbf{k}}(\mathbf{r}) = e^{i\mathbf{k}\mathbf{r}} + \int d^3r' G(\mathbf{r}-\mathbf{r}')V(\mathbf{r}')\psi_{\mathbf{k}}(\mathbf{r}'),$$

that is, the Green's function term, has to be a sum over *outgoing waves only*. And, we can guarantee this by distorting the contour of integration in the right direction, as follows.

The contour integral has to be evaluated by closing the contour. Since r is positive $e^{ik'r}$ goes to zero in the *upper* half k¢ plane, but diverges in the lower half, so we must close the contour in the upper half plane to ensure no contribution from the semicircle at infinity. Therefore, to get the desired outgoing waves, e^{ikr} but not e^{-ikr}, our contour closed in the upper half plane must encircle the pole at $k' = +k$ but *not* the one at $k' = -k$. (e^{ikr} *does* represent outgoing waves: the suppressed time dependence is $e^{-iEt/\hbar} = e^{-iwt}$, giving $e^{i(kr - wt)}$.) In other words, the relative configuration of the real-axis part of the contour and the two poles has to be:

x (pole)

x (pole at $k' = -k - i\varepsilon$)

(pole at $k' = +k + i\varepsilon$)

Instead of moving the contour slightly off the real axis to avoid the poles, we've moved the *poles* slightly instead. These movements are infinitesimal, so which gets moved makes no difference to the value of the integral. It is more convenient to move the poles, as shown, because this move can be efficiently included in the integral just by adding an infinitesimal imaginary part to the denominator:

$$G_+(r,k) = \int \frac{d^3k'}{(2\pi)^3} \frac{e^{ik'r}}{E_k - \frac{\hbar^2 k'^2}{2m} + i\varepsilon} = -\frac{m}{2\pi^2 ir\hbar^2} \int_{-\infty}^{\infty} \frac{k'dk'e^{ik'r}}{k'^2 - k^2 - i\varepsilon}$$

Notice that we have written G_+ instead of G, because G can denote any solution of

$$\left(\frac{\hbar^2}{2m}\nabla^2 + E_k\right) G(\mathbf{r},\mathbf{k}) = \delta(\mathbf{r})$$

and we are specifying the particular solution having only *outgoing* waves. *In contrast to G, G_+ is well-defined and unique.* (There is another perfectly valid solution having only ingoing waves, but it is irrelevant to the scattering problem. The difference between the ingoing and outgoing solutions satisfies the homogeneous equation having zero on the right-hand side.)

Once we move the poles slightly as described above, the pole at k¢ = +k + *ie* is in fact the *only* singularity of the integrand lying inside the contour of integration (closed in the upper half plane), so the value of the integral is just the contribution from this pole, that is,

$$G_+(r,k) = -\frac{m}{2\pi^2 ir\hbar^2}(2\pi i)\frac{ke^{ikr}}{2k} = -\frac{m}{2\pi\hbar^2}\frac{e^{ikr}}{r}.$$

Therefore the *ie* prescription (as it's sometimes called) in G_+ does indeed give us what we want: a solution having *only outgoing waves,* and the integral equation becomes:

$$\psi_{\mathbf{k}}(\mathbf{r}) = e^{i\mathbf{k}\cdot\mathbf{r}} - \frac{m}{2\pi\hbar^2}\int d^3r' \frac{e^{ik|\mathbf{r}-\mathbf{r}'|}}{|\mathbf{r}-\mathbf{r}'|} V(\mathbf{r}')\psi_{\mathbf{k}}(\mathbf{r}').$$

This can be written more simply if we assume the potential to be localized, so that we can take $|r| >> |r'|$. In this case, it is a good approximation to take $|r - r'| = r$ in the denominator. However, this approximation cannot be made in the exponential, because to leading order

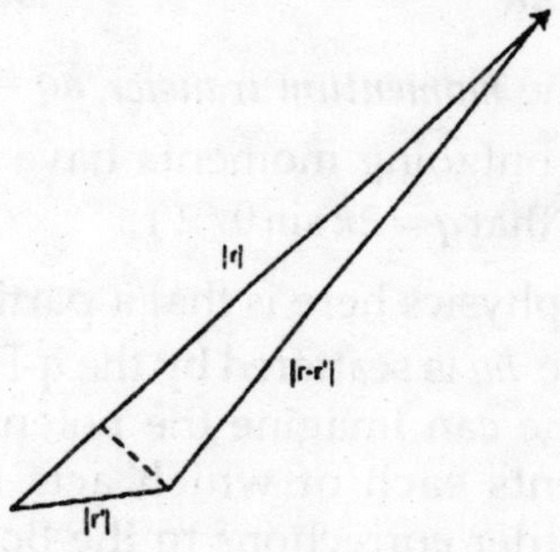

$$k|r - r'| = kr - k\hat{r}.r' = kr - k_f.r'$$

and although the second term is much smaller than the first, it is a *phase,* which may be of order unity. Such a factor must of course be included so that the contributions to the integral from different regions of the potential are added with the correct relative phases.

Therefore, assuming the detector distance r is much larger than the range of the potential, we can write

$$\psi_{\mathbf{k}}(\mathbf{r}) = e^{i\mathbf{k}\mathbf{r}} - \frac{m}{2\pi\hbar^2}\frac{e^{ikr}}{r}\int d^3r' e^{-i\mathbf{k}_f\mathbf{r}'} V(\mathbf{r}')\psi_{\mathbf{k}}(\mathbf{r}').$$

THE BORN APPROXIMATION

From the above equation, the first order approximation to the scattering is given by replacing ψ in the integral on the right with the zeroth-order term $e^{ik.r}$,

$$\psi_{\mathbf{k}(Born)}(\mathbf{r}) = e^{i\mathbf{k}\mathbf{r}} - \frac{m}{2\pi\hbar^2}\frac{e^{ikr}}{r}\int d^3r' e^{i\mathbf{k}\mathbf{r}' - i\mathbf{k}_f\mathbf{r}'} V(\mathbf{r}'),$$

This is the *Born approximation*. In terms of the scattering amplitude *f(q, j),* which we defined in terms of the asymptotic wave function:

$$\psi_{\mathbf{k}}(\mathbf{r}) = e^{i\mathbf{k}\mathbf{r}} + f(\theta,\varphi)\frac{e^{ikr}}{r}$$

the Born approximation is:

$$f_{\text{Born}}(\theta,\varphi) = -\frac{m}{2\pi\hbar^2}\int d^3r' e^{i\mathbf{k}\mathbf{r}'-\mathbf{k}_f\mathbf{r}'}V(\mathbf{r}') = -\frac{m}{2\pi\hbar^2}\int d^3r' e^{-i\mathbf{q}\mathbf{r}'}V(\mathbf{r}')$$

where $\hbar q$ is the *momentum transfer*, $\hbar q = \hbar\left(k_f - k\right)$. (Since the incoming and outgoing momenta have equal magnitude, it is easy to check that $q = 2k\sin\theta/2$)

The essential physics here is that a particle scattered with momentum change $\hbar q$ is scattered by the q-Fourier component of the potential one can imagine the potential as built up of Fourier components each of which acts like a diffraction grating. Higher order corrections to the Born approximation correspond to successive scatterings off these gratings these higher orders are generated by iteration of

$$\psi_{\mathbf{k}}(\mathbf{r}) = e^{i\mathbf{k}\mathbf{r}} - \frac{m}{2\pi\hbar^2}\int d^3r' \frac{e^{ik|\mathbf{r}-\mathbf{r}'|}}{|\mathbf{r}-\mathbf{r}'|}V(\mathbf{r}')\psi_{\mathbf{k}}(\mathbf{r}').$$

It is important to establish when the Born approximation is a good one: sometimes it isn't. Actually, we are just doing perturbation theory in disguise, so we need the perturbation to be small, that is to say, replacing $y_k(r')$ by $e^{ik.r2}$ in the integral on the right in the equation above should only make a small difference to the value of $y_k(r)$ given by doing the integral. This is of course a rather tricky exercise in self-consistency.

Let us attempt to estimate what difference the replacement of $y_k(r')$ by $e^{ik.r'}$ in the integrand *does* make for the common case of a spherically symmetric potential *V*(*r*) parameterized by depth V_0 and range r_0. The integral is effectively only over a region of size r_0 around the origin.

First consider *low energy scattering*, $kr_0 < 1$ say, so for estimation purposes we can replace the exponential term by 1 in the region of integration. We also assume that where y appears in the integral on the right-hand side of the equation $\left|\psi_k(r')\right|$ is also pretty close to 1 (remember the integral is only over a volume within r_0 or so of the origin) and so we just replace it by 1. In other words, we're assuming that the ingoing plane wave, the $e^{i\mathbf{k}.\mathbf{r}}$, is not dramatically distorted inside that volume where the potential is significant.

Now, we've assumed the wave function near the origin is close to 1, so putting that value in the integrand on the right had better give a value for $\psi_k(r)$ on the left hand side of the equation which is pretty close to 1. The approximations give:

$$\psi_{\mathbf{k}}(0) \approx 1 - \frac{m}{2\pi\hbar^2}\int d^3r' \frac{V(r')}{r'},$$

so the Born approximation will be reasonable at low energies ($kr_0 < 1$) *if* the second term on the right hand side is a lot less than unity.

When is this true for a real potential? Taking $V(r)$ to have depth V_0 and range r_0, the Born approximation is good if:

$$\frac{m}{2\pi\hbar^2}\int_0^{r_0} 4\pi r^2 \frac{V_0}{r} dr << 1, \text{ or } V_0 << \frac{\hbar^2}{mr_0^2}.$$

Notice that the right hand side of this inequality is of order the kinetic energy of a particle confined to a volume equal to the range of the potential, so *the Born approximation is valid at low energies provided the potential is well below the strength necessary for a bound state.* In fact, the Born approximation works better at higher energies, because the oscillating phase term in $-\dfrac{m}{2\pi\hbar^2}\int d^3r' e^{-iqrV(r')}$ cuts down the value of the integral by a factor of order of magnitude $1/(kr_0)$. This means the condition becomes $V_0 << kr_0 \dfrac{\hbar^2}{mr_0^2}$ always satisfied at high enough energies.

THE LIPPMANN-SCHWINGER EQUATION

It proves illuminating, especially in understanding scattering beyond the Born approximation, to recast the Green's function derivation of the scattering amplitude in the more formal language of bras, kets and operators. The Green's function was introduced in the previous section as the (non-unique) inverse of the operator

$$E_k - H_0 = \left(\frac{\hbar^2}{2m} \nabla^2 + E_k \right).$$

(Parenthetical remark: in numerical computation, the wavefunction might be specified at points on a lattice in space, and a differential operator like this would be represented as a difference operator, that is, as a large but finite matrix operating on a large vector whose elements were the wavefunction values at points on the lattice. The Green's function would then be the inverse matrix with appropriate boundary conditions specified to ensure uniqueness.)

Purely formally (and following Sakurai), writing $H = H_0 + V$, with H_0 the kinetic energy operator $p^2/2m$, the ingoing plane wave state is a solution of

$$H_0 \,|\, k >= E_k \,|\, k >$$

We want to solve

$$(H_0 + V) \,|\, \psi >= E_k \,|\, \psi >$$

The transformation from a differential equation to an integral equation in this language is:

$$|\, \psi >= |\, \mathbf{k} > + \frac{1}{E_k - H_0} V \,|\, \psi >$$

This gives the undisturbed incoming wave for $V = 0$, and by operating on both sides of the equation with $E - H_0$, we find $|\psi>$ does indeed satisfy the full Schrödinger equation. But of course this transformation from a differential to an integral equation has the same flaw as the earlier treatment: H_0 has a continuum of eigenvalues in the infinite volume limit, so the operator equation becomes ill-defined for those eigenstates with energy arbitrarily close to the incoming energy, and those are precisely the states of physical relevance.

To make explicit that this is indeed the problem we've already solved, let us translate it into the earlier language. First take the inner product with the bra <r|:

$$<\mathbf{r}|\psi> = <\mathbf{r}|\mathbf{k}> + <\mathbf{r}|\frac{1}{E_k - H_0}V|\psi>.$$

Insert a representation of unity as a sum over eigenstates of momentum (and therefore of H_0) into the last term:

$$<\mathbf{r}|\psi> = <\mathbf{r}|\mathbf{k}> + \int\frac{d^3k'}{(2\pi)^3}<\mathbf{r}|\mathbf{k}'><\mathbf{k}'|\frac{1}{E_k - H_0}V|\psi>$$

$$= <\mathbf{r}|\mathbf{k}> + \int\frac{d^3k'}{(2\pi)^3}<\mathbf{r}|\mathbf{k}'>\frac{1}{E_k - E_{k'}}<\mathbf{k}'|V|\psi>.$$

Finally, insert *another* representation of unity as a sum over eigenstates of *position* in the last term:

$$<\mathbf{r}|\psi> = <\mathbf{r}|\mathbf{k}> + \int d^3r'\int\frac{d^3k'}{(2\pi)^3}<\mathbf{r}|\mathbf{k}'>\frac{1}{E_k - E_{k'}}<\mathbf{k}'|\mathbf{r}'><\mathbf{r}'|V|\psi>.$$

Comparing this expression with the integral equation in the earlier discussion, it is evident that they are indeed equivalent, and therefore the correct $i\varepsilon$ prescription to give the scattered wave function,

$$|\psi> = |\mathbf{k}> + \frac{1}{E_k - H_0 + i\varepsilon}V|\psi> = |\mathbf{k}> + G_+V|\psi>$$

where

$$G_+ = \frac{1}{E_k - H_0 + i\varepsilon} = \int\frac{d^3k'}{(2\pi)^3}\frac{|k'><k'|}{E_k - E_{k'} + i\varepsilon}$$

in which form it is evident that $<r|G_+|r'>$ is the same as $G_+(r - r')$ in the previous work.

This equation for the scattered wave |y> is called the Lippmann-Schwinger equation.

Now that we have a well-defined Green's function operator G_+, the Lippmann-Schwinger equation can be solved formally:

$$|\psi> = |\mathbf{k}> + G_+V|\psi>, \text{ so } |\psi> = \frac{1}{1 - G_+V}|\mathbf{k}>,$$

with a series solution $|\psi> = |k> + G_+V|k> + G_+VG_+V|k>$ $+G_+VG_+VG_+V|k> ...$ just a formal version of the solution we found earlier.

THE TRANSITION MATRIX

Operating on both sides of the above the equation with V,

$$V|\psi> = V|k> + VG_+V|k> + VG_+VG_+V|k> ... = T|k>$$

defining the "transition matrix" T by

$$T = V + VG_+V + VG_+VG_+V + ... = V + V\frac{1}{E - H_0 + i\varepsilon}V + ...$$

In terms of this transition matrix operator, the scattered wave can be written $|\psi> = |k> + G_+T|k>$ Comparing this with,and recalling that the Born approximation is given by $|\psi> = |k> + G_+T|\psi>$,we see that T is a kind of generalized potential, including all the higher order terms, so that just as the Born approximation gave the scattering amplitude in terms of V,

$$f_{Born}(\theta,\varphi) = -\frac{m}{2\pi\hbar^2}\int d^3r' e^{ikr' - k_i r'} V(r')$$

the *exact* result including *all* higher order terms must have the same structure with T replacing V. Of course, unlike V(r), T is not a diagonal matrix in r-space: it depends on two space variables, and its Fourier transform is a therefore function of two momenta, that is, the incoming k and the scattered k′. Thus we find:

$$f(\theta,\varphi) = -\frac{m}{2\pi\hbar^2}\int d^3r\int d^3r' e^{i\mathbf{kr} - i\mathbf{k'r'}} T(\mathbf{r}',\mathbf{r}) = -\frac{m}{2\pi\hbar^2} < \mathbf{k}'|T|\mathbf{k} >.$$

We have replaced the k_f in the Born expression with k′. Sakurai has an extra $(2p)^3$ in the term on the right, because he uses $\langle\vec{k}|\vec{k}'\rangle = \delta(\vec{k} - \vec{k}')$, $\langle\vec{k}|\vec{k}\rangle = \frac{e^{i\vec{k}.\vec{r}}}{(2\pi)^{3/2}}$, we use $\langle\vec{k}|\vec{k}'\rangle = (2\pi)^3\delta(\vec{k} - \vec{k}')$, $\langle\vec{r}|\vec{k}\rangle = e^{i\vec{k}.\vec{r}}$,.

THE OPTICAL THEOREM

The Optical Theorem relates the imaginary part of the forward scattering amplitude to the total cross-section,

$$\operatorname{Im} f(\theta = 0) = \frac{k\sigma_{tot}}{4\pi}.$$

The physical content of this initially mysterious theorem will become a lot clearer after we discuss partial waves and some geometric effects. It *does* tell us that *f* cannot be real in all directions, and that in particular *f* has a positive imaginary part in the forward direction. We've included the proof here for the record, but you can skip it for now. But note that this proof is more general than the simple one given (later) in the section on partial waves, in that we do not here assume the potential to have spherical symmetry.

From the expression for $f(\theta, \varphi)$ above, we see that we must find the imaginary part of $<k|T|k>$.

Recall that

$$V|\psi> = T|\mathbf{k}>,$$

so we need to find

$$\operatorname{Im} <\mathbf{k}|V|\psi> = \operatorname{Im}\left[\left(<\psi| - <\psi|V\frac{1}{E - H_0 - i\varepsilon}\right)V|\psi>\right].$$

Since *V* is hermitian, the only imaginary part of the above matrix element comes from the *ie*, recalling that

$$\frac{1}{E - H_0 - i\varepsilon} = \frac{P}{E - H_0} + i\pi\delta(E - H_0).$$

Therefore,

$$\operatorname{Im} <\mathbf{k}|V|\psi> = \pi <\psi|V\delta(E - H_0)V|\psi>$$

Again using

$$V|\psi> = T|\mathbf{k}>$$

we can rewrite the equation

$$\operatorname{Im} <\mathbf{k}|T|\mathbf{k}> = \operatorname{Im} <\mathbf{k}|V|\psi> = -\pi <\psi|V\delta(E - H_0)V|\psi> =$$

$$-\pi <\mathbf{k}\,|\,T^{\dagger}\,\delta(E-H_0)T\,|\,\mathbf{k}>.$$

Inserting a complete set of plane wave states in the final matrix element above gives

$$\begin{aligned}\operatorname{Im} <\mathbf{k}\,|\,T\,|\,\mathbf{k}> &= -\pi <\mathbf{k}\,|\,T^{\dagger}\delta(E-H_0)T\,|\,\mathbf{k}> \\ &= -\pi\int\frac{d^3k'}{(2\pi)^3} <\mathbf{k}\,|\,T^{\dagger}\,|\,\mathbf{k}'><\mathbf{k}'\,|\,T\,|\,\mathbf{k}>\delta\left(E-\frac{\hbar^2k'^2}{2m}\right) \\ &= -\pi\int\frac{d\Omega'}{(2\pi)^3}\frac{mk}{\hbar^2}|<\mathbf{k}'\,|\,T\,|\,\mathbf{k}>|^2.\end{aligned}$$

TIME-DEPENDENT FORMULATION OF SCATTERING THEORY

In the time-independent formulation presented above, we solved the Lippmann-Schwinger equation to find

$$|\psi>=|\mathbf{k}>+G_+V\,|\,\mathbf{k}>+G_+VG_+V\,|\,\mathbf{k}>+G_+VG_+VG_+V\,|\,\mathbf{k}>\ldots$$

where $G_+(E)=\dfrac{1}{E-H_0+i\varepsilon}=\displaystyle\int\frac{d^3k'}{(2\pi)^3}\frac{|\vec{k}'><\vec{k}'|}{E-E_{k'}+i\varepsilon}$

and $E = E_k$.

(*Reminder on our wave function normalization convention*: we always have a denominator 2p for an integral *dk*. This means the identity operator as a sum over plane wave projection operators is $I=\displaystyle\int\frac{d^3l}{(2\pi)^3}|\,\vec{k}\rangle\langle\vec{k}\,|$. To see how this comes about, take the matrix element between two position eigenstates and Fourier transform from energy to time:

$$\begin{aligned}G_+(\vec{r},\vec{r}',t) &= \frac{1}{2\pi}\int e^{-iEt/\hbar}dE\langle\vec{r}\,|G_+|\,\vec{r}'\rangle \\ &= \frac{1}{2\pi}\int e^{-iEt/\hbar}dE\int\frac{d^3k'}{(2\pi)^3}\frac{\langle\vec{r}\,|\,k'\rangle\langle k'\,|\,\vec{r}'\rangle}{E-E_{k'}+i\varepsilon} \\ &= \frac{1}{2\pi}\int e^{-iEt/\hbar}dE\int\frac{d^3k'}{(2\pi)^3}\frac{e^{i\vec{k}'\cdot(\vec{r}-\vec{r}')}}{E-E_{k'}+i\varepsilon}.\end{aligned}$$

The integral over *E* is along the real axis, and the contour is closed in the half plane where the integrand goes to zero for in the imaginary direction, that is, in the lower half plane for $t > 0$ and the upper half plane for $t < 0$. But with the $i\varepsilon$ term shown, all the singularities of the integrand are in the *lower* half plane. *Hence* G_+ *is identically zero for* $t < 0$.

For $t > 0$, G_+ is just the free particle propagator between the two points (apart from the phase factor $-i$):

$$G_+(\vec{r},\vec{r}\,',t) = -i\int\frac{d^3k'}{(2\pi)^3} i\bar{k}(\bar{r}-\bar{r}) - iE_h t/\hbar$$

To summarize: terms in the series solution of the Lippmann-Schwinger equation can be interpreted as successive scatterings off the Fourier components of a potential, with plane wave propagation in between, with the sign of the $i\varepsilon$ term ensuring that there are only outgoing waves from each scattering. In the Fourier transformed version above, the sum is over scattering at all possible points where the potential is nonzero, with G_+ propagation in between, the $i\varepsilon$ ensuring that the scattering path only moves forward in time.

Last semester, we defined the free-particle propagator as the operator $U(t) = e^{-H_0 t/\hbar}$ The propagator describes development of the free-particle wave function in time, so naturally $U(t) = 0 \text{ for } t < 0$ Then Fourier transforming the propagator from t to E, and inserting an infinitesimal exponentially decaying factor to define the integral at infinity, we find

$$U(E) = \int_0^\infty e^{iEt/\hbar} e^{-iH_0 t/\hbar} e^{-\varepsilon t/\hbar} dt = \frac{i\hbar}{E - H_0 + i\varepsilon}.$$

Note that the propagators U and G_+ differ by a factor of $i\hbar$, specifically

$$G_+(t) = -\frac{i}{\hbar}\theta(t) e^{-H_0 t/\hbar}$$

This is the correctly normalized Green's function for the time-dependent free-particle Schrödinger equation: it is the solution of

$$\left(i\hbar\frac{\partial}{\partial t}-H_0\right)G_+(t)=\delta(t)$$

which propagates forwards in time. The reason the propagators U and G_+ differ by a factor $i\hbar$ of is that the Lippmann-Schwinger equation can be generated as a time sequence using the higher-order perturbation theory interaction representation described earlier in the course, essentially expanding $e^{-i(H_0+V)t/\hbar}$ as a time-ordered series expansion in V, and each factor V has an accompanying $1/(i\hbar)$, these factors are taken care of by using G_+ instead of U.

Exercise: in that earlier lecture, we gave the second-order term as:

$$c_n^{(2)}(t)=\left(\frac{1}{i\hbar}\right)^2\sum_n\int_0^t\int_0^{t'}dt'dt''e^{-i\omega_f(t-t')}<f|V_S(t')|n>e^{-i\omega_n(t'-t'')}<n|V_S(t'')|i>e^{-i\omega_i t''}$$

Assume that the potential V is constant in time. Fourier transform this expression from t to E ($E=\hbar\omega$), and establish that it has the structure $G_+(E)VG_+(E)VG_+(E)$.

THE BORN CROSS-SECTION FROM TIME-DEPENDENT THEORY

We established in the lecture on Time-Dependent Perturbation Theory that to leading order in the perturbation, the transition rate from an initial state i to a final state f is given by Fermi's Golden Rule:

$$R_{i\to f}=\frac{2\pi}{\hbar}\left|\langle f|V|i\rangle\right|^2\delta\left(E_f-E_i\right).$$

We can use this result to find in leading order the rate of scattering from an incoming plane wave into any outgoing plane wave state having the same energy, and hence by adding the rate over the plane wave directions pointing within a given small solid angle $d\Omega$, rederive the Born approximation.

Conceptually, though, this is a bit tricky. From the above solution of the Schrödinger equation, we know the outgoing wave is a spherical one, so in a particular direction the amplitude decreases. But that doesn't happen with a plane wave! The clearest way to handle this is to put the system in a big box, a cube of side *L*, with periodic boundary conditions. This makes it easier to count states and normalize the plane waves properly of course, in the limit of a large box, the plane waves form a complete set, so any spherical wave *can* be expressed as a sum over these plane waves.

In this section, then, we use box-normalized plane waves:

$$\left|\vec{k}\right\rangle = \frac{1}{L^{3/2}} e^{i\vec{k}.\vec{r}}, \quad \left\langle\vec{k}'\middle|\vec{k}\right\rangle = \delta_{\vec{k}\vec{k}'} .$$

So

$$\left\langle f\middle|V\middle|i\right\rangle = \frac{1}{L^3}\int d^3r e^{-i\vec{k}_f.\vec{r}} V(\vec{r}) e^{i\vec{k}_i.\vec{r}} = \frac{1}{L^3}\int d^3r e^{-i\vec{q}.\vec{r}} V(\vec{r})$$

where the momentum transfer to the particle $\hbar\vec{q} = \hbar\left(\vec{k}_f - \vec{k}_i\right)$

It is important to note that we are taking the incoming wave to be just *one* of the normalized plane wave states satisfying the box periodic boundary conditions, so now the incoming current, being from just one of these plane waves, is

$$j_{in} = |\psi|^2 v = \frac{1}{L^3}\frac{p}{m}$$

The Golden Rule becomes

$$R_{i\to(f \text{ in } d\Omega)} = \frac{2\pi}{\hbar}\left|\left\langle f\middle|V\middle|i\right\rangle\right|^2 \delta\left(E_f - E_i\right) d\Omega$$

f denoting a plane wave going outwards within the solid angle $d\Omega$.

Now, the *d*-function simply counts the number of states available at the correct (initial) energy, within the specified final solid angle of direction. The density of states in momentum space (for volume L^3 of real space) is one state in each momentum-space volume $(2\pi\hbar)^3 / L^3$, so using $dE/dp = p/m$, the density of states in *energy* for outgoing solid angle $d\Omega$ is $L^3 mpd\Omega/(2\pi\hbar)^3$.

Putting this all together

$$R_{i\to(f\text{ in }d\Omega)}=\frac{2\pi}{\hbar}\left|\frac{1}{L^3}\int d^3x e^{-i\vec{q}.\vec{x}}V(\vec{x})\right|^2 L^3 mpd\Omega/(2\pi\hbar)^3.$$

The transition rate, the rate of scattering into $d\Omega$, is just the incident current multiplied by the infinitesimal scattering cross-section $d\sigma$ (q, φ) (that was our definition of $d\sigma$),

$$j_{in}\left(\frac{d\sigma(\theta,\varphi)}{d\Omega}\right)d\Omega=R_{i\to(f\text{ in }d\Omega)}$$

because our definition of $R_{i\to(f\text{ in }d\Omega)}$ included the appropriately normalized ingoing wave.

so finally

$$\frac{d\sigma(\theta,\varphi)}{d\Omega}=\frac{R_{i\to(f\text{ in }d\Omega)}}{j_{in}d\Omega}=\frac{m}{p}\frac{2\pi}{\hbar}\left|\int d^3re^{-i\vec{q}.\vec{r}}V(\vec{r})\right|^2 mp/(2\pi\hbar)^3=\left|\frac{m}{2\pi\hbar^2}\int d^3re^{-i\vec{q}.\vec{r}}V(\vec{r})\right|^2$$

$$R_{i\to(f\text{ in }d\Omega)}=\frac{2\pi}{\hbar}\left|\langle f|V|i\rangle\right|^2\delta(E_f-E_i)d\Omega$$

In the continuum version, $\langle\vec{r}\,|\,\vec{k}\rangle=e^{i\vec{k}.\vec{r}}$, so the matrix element term is just $\left|\int d^3re^{-i\vec{q}.\vec{r}}V(\vec{r})\right|^2$ The energy *d*-function is only meaningful inside an integral, in this case over the small volume of outgoing scattering states in the solid angle dW and energy equal to the ingoing energy. But this integral over *k*2 - space *must* include the $1/(2p)^3$ factor, according to our rule, giving an outgoing phase space term

$$\int\frac{d^3k'}{(2\pi)^3}\delta(E_{k'}-E_k)=d\Omega\int\frac{k'^2dk'}{(2\pi)^3}\delta\left(\frac{\hbar^2k'^2}{2m}-\frac{\hbar^2k^2}{2m}\right)=d\Omega\frac{k^2}{(2\pi)^3}\frac{m}{\hbar^2k}=d\Omega\frac{mp}{(2\pi\hbar)^3}.$$

This establishes that our continuum normalization conventions give the same result as that obtained from box normalization.

ELECTRONS SCATTERING FROM ATOMS

This same approach, using the Golden Rule to derive the leading order scattering rate, is useful is analyzing the scattering of fast electrons by atoms. The problem with slow electrons is that the wave function needs to be antisymmetric with respect to all electrons present. We assume fast electrons have little overlap with the atomic electron wave functions in momentum space, so we don't have to worry about symmetry.

With this approximation, the scattering amplitude matrix element is

$$\int d^3r e^{i\vec{q}.\vec{r}} \langle n|\left(-\frac{Ze^2}{r}+\sum_i \frac{e^2}{|\vec{r}-\vec{r}_i|}\right)|0\rangle$$

where the potential term $V(\vec{x})$ is that from the nucleus, plus the repulsion from the other electrons at positions $\vec{x}_i$. Taking the final atomic state n allows for the possibility of inelastic scattering.

Since the distance r of the scattered electron from the nucleus has nothing to do with the atomic state, $n = 0$ for the nuclear contribution, which is then just Coulomb scattering, and

$$\int d^3r \frac{e^{i\vec{q}.\vec{r}}}{r} = \frac{4\pi}{q^2}.$$

(To do this integral, put in a convergence factor $e^{-\varepsilon r}$ then let $\varepsilon \to 0$)

The term involving the atomic electrons is another matter: for the ith electron, integrating over the coordinate of the scattered electron gives a factor $\left(4\pi/q^2\right)e^{ith.\xi_1}$, but the hard part is finding the value of the matrix element of this operator between the atomic states. Notice that this is just the Fourier transform of the electrostatic potential from the ith electron's charge density,

$\nabla^2 V_i(\vec{r}_i) = 4\pi\rho_i(\vec{r}_i)$ transforms to

$V_i(\vec{q}) = \left(4\pi/q^2\right)\rho_i(\vec{q})$ and $\rho_i(\vec{r}) = e\delta(\vec{r}-\vec{r}_i)$ Fourier transforms to $e\left(e^{i2f\zeta_1}\right)$.

THE FORM FACTOR

For elastic scattering, then, the contribution of the atomic electrons is simply interpreted: their charge density gives rise to a potential by the usual electrostatic equation, and the (fast) electron is scattered by this potential. For inelastic scattering, the Fourier transform of the electron density is evaluated between different atomic states. In both cases, the matrix element is called the *form factor* $F_n(\vec{q})$ for the scattering, actually $ZF_n(\vec{q}) = \langle n | \sum_i e^{i\vec{q}\vec{r}_i} | 0 \rangle$. The normalizing factor Z is introduced so that for elastic scattering, $F_n(\vec{q}) \to 1$ as $q \to 0$.

So the form factor is a map of the charge density in q-space. By measuring the scattering rate at different angles, and Fourier analyzing, it is possible to delineate the charge distribution in ordinary space. The same technique works for nuclei, and in fact for particles the neutron, for example, although electrically neutral, has a nontrivial electrical charge distribution within its volume, revealed by scattering very fast electrons.

More general form factors describe distribution of spin, and also time dependence of distributions of charge or spin in excited systems. These can all be measured with suitably designed scattering experiments.

Chapter 13

More Scattering

We are considering the solution to Schrödinger's equation for scattering of an incoming plane wave in the *z*-direction by a potential localized in a region near the origin, so that the total wave function beyond the range of the potential has the form

$$\psi(r,\theta,\varphi) = e^{ikr\cos\theta} + f(\theta,\varphi)\frac{e^{ikr}}{r}.$$

The overall normalization is of no concern, we are only interested in the *fraction* of the ingoing wave that is scattered. Clearly the outgoing current generated by scattering into a solid angle dΩ at angle q, φ is $|f(\theta,\varphi)|^2 \, d\Omega$ multiplied by a velocity factor that also appears in the incoming wave.

Many potentials in nature are spherically symmetric, or nearly so, and from a theorist's point of view it would be nice if the experimentalists could exploit this symmetry by arranging to send in spherical waves corresponding to different angular momenta rather than breaking the symmetry by choosing a particular direction. Unfortunately, this is difficult to arrange, and we must be satisfied with the remaining azimuthal symmetry of rotations about the ingoing beam direction.

In fact, though, a full analysis of the outgoing scattered waves from an ingoing plane wave yields the same information as would spherical wave scattering. This is because a plane wave can actually be written as a *sum over spherical waves*:

$$e^{i\vec{k}.\vec{r}} = e^{ikr\cos\theta} = \sum_l i^l (2l+1) j_l(kr) P_l(\cos\theta)$$

Visualizing this plane wave flowing past the origin, it is clear that in spherical terms the plane wave contains both incoming *and* outgoing spherical waves. The real function $j_i(kr)$ is a standing wave, made up of incoming and outgoing waves of equal amplitude.

We are, obviously, interested only in the outgoing spherical waves *that originate by scattering from the potential,* so we must be careful not to confuse the pre-existing outgoing wave components of the plane wave with the *new* outgoing waves generated by the potential.

The radial functions $j_i(kr)$ appearing in the above expansion of a plane wave in its spherical components are the *spherical Bessel functions,* discussed below. The azimuthal rotational symmetry of plane wave + spherical potential around the direction of the ingoing wave ensures that the angular dependence of the wave function is just $p_i(\cos\theta)$, not $Y_{lm}(\theta,\varphi)$. The coefficient $i^l(2l+1)$ is derived in Landau and Lifshitz, §34, by comparing the coefficient of $(kr\cos\theta)^n$ on the two sides of the equation: $(kr)^n$ does not appear in $j_l(kr)$ for l greater than n, and $(\cos\theta)^n$ does not appear $P_l(\cos\theta)$ in $(kr\cos\theta)^n$ for l less than n, so the *combination* only occurs once in the nth term, and the coefficients on both sides of the equation can be matched.

THE SPHERICAL BESSEL AND NEUMANN FUNCTIONS

The plane wave $e^{i\vec{k}\vec{r}}$ is a trivial solution of Schrödinger's equation with zero potential, and therefore, since the $P_l(\cos\theta)$ form a linearly independent set, each term

$j_l(kr)P_l(\cos\theta)$ in the plane wave series must be itself a solution to the zero-potential Schrödinger's equation. It follows that satisfies $j_l(kr)$ the zero-potential radial Schrödinger equation:

$$\frac{d^2}{dr^2}R_l(r)+\frac{2}{r}\frac{d}{dr}R_l(r)+\left(k^2-\frac{l(l+1)}{r^2}\right)R_l(r)=0.$$

The standard substitution $R_l(r)=u_l(r)/r$ yields

$$\frac{d^2u_l(r)}{dr^2}+\left(k^2-\frac{l(l+1)}{r^2}\right)u(r)=0$$

For the simple case $l = 0$ the two solutions are $u_0(r)=\sin kr, \cos kr$. The corresponding radial functions $R_0(r)$ are (apart from overall constants) the zeroth-order *Bessel* and *Neumann* functions respectively.

The standard normalization for the zeroth-order Bessel function is $j_0(kr)=\frac{\sin kr}{kr}$

and the zeroth-order Neumann function $n_0(kr)=-\frac{\cos kr}{kr}$

Note that the Bessel function is the one well-behaved at the origin: it could be generated by integrating out from the origin with initial boundary conditions of value one, slope zero.

Here is a plot of $j_0(kr)$ and $n_0(kr)$ from $kr = 0.1$ to 20:

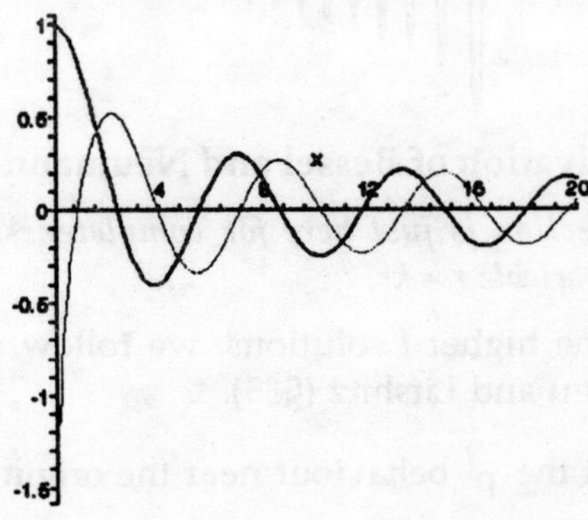

For nonzero l, near the origin is $R_l(r) \sim r^l$ or $\sim r^{-(l+1)}$ The well-behaved $\sim r^l$ solution is the Bessel function, the singular function the Neumann function. The standard normalizations of these functions are given below.

Here are $j_s(kr)$ and $j_{50}(kr)$:

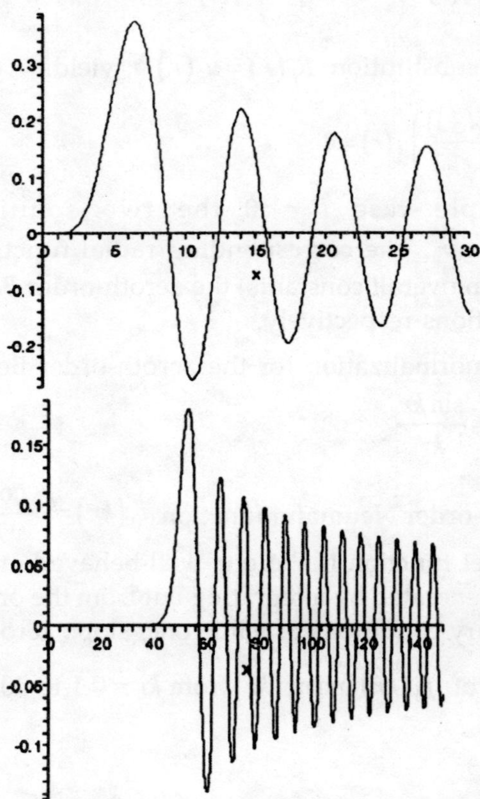

Detailed Derivation of Bessel and Neumann Functions

This subsection is just here for completeness. We use the dimensionless variable r = kr.

To find the higher l solutions, we follow a clever trick given in Landau and Lifshitz (§33).

Factor out the ρ^l behaviour near the origin by writing

The function satisfies

$$R_l = (\rho)^l \chi_l(\rho).$$

The trick is to differentiate this equation with respect to *r*:

$$\frac{d^2}{d\rho^2}\chi_l(\rho)+\frac{2(l+1)}{\rho}\frac{d}{d\rho}\chi_l(\rho)+\chi_l(\rho)=0.$$

Writing purely formally $\frac{d}{d\rho}\chi(\rho)=\rho\chi_{l+1}(\rho)$, the equation becomes

$$\frac{d^3}{d\rho^3}\chi_l(\rho)+\frac{2(l+1)}{\rho}\frac{d^2}{d\rho^2}\chi_l(\rho)+\left(1-\frac{2(l+1)}{\rho^2}\right)\frac{d}{d\rho}\chi_l(\rho)=0.$$

But this *is* the equation that $\chi_{l+1}(\rho)$ must obey! So we have a *recursion formula* for generating *all* the $j_i(\rho)$ from the zeroth one: $\chi_{l+1}(\rho)=\frac{1}{\rho}\frac{d}{d\rho}\chi_l(\rho)$ and $j_l(\rho)=(\rho)^l\chi_l(\rho)$, up to a normalization constant fixed by convention.

In fact, the standard normalization is

$$j_l(\rho)=(-\rho)^l\left(\frac{1}{\rho}\frac{d}{d\rho}\right)^l\left(\frac{\sin\rho}{\rho}\right).$$

Now $(\sin\rho)/\rho=\sum_0^\infty(-1)^n\rho^{2n/(2n+1)!}$. This is a sum of only *even* powers of ρ. It is easily checked that operating on this series with $\left(\frac{1}{\rho}\frac{d}{d\rho}\right)^l$ can never generate any negative powers of *r*. It follows that $j_l(\rho)$ written as a power series in *r*, has leading term proportional to ρ^l. The coefficient of this leading term can be found by applying the differential operator to the series for $(\sin\rho)/\rho$,

$$j_l(\rho)\sim\frac{\rho^l}{(2l+1)!!}\quad\text{as } \rho\to 0.$$

This r^l behaviour near the origin is the usual well-behaved solution to Schrödinger's equation in the region where the centrifugal term dominates.

Note that the small ρ behaviour is *not* immediately evident from the usual presentation of the $j_l(\rho)$'s, written as a mix of powers and trigonometric functions, for example

$$j_1(\rho)=\frac{\sin\rho}{\rho^2}-\frac{\cos\rho}{\rho},\quad j_2(\rho)=\left(\frac{3}{\rho^3}-\frac{1}{\rho}\right)\sin\rho-\frac{3\cos\rho}{\rho^2},\quad \text{etc.}$$

Turning now to the behaviour of the $j_l(\rho)$'s for *large r*, from

$$j_l(\rho)=(-\rho)^l\left(\frac{1}{\rho}\frac{d}{d\rho}\right)^l\left(\frac{\sin\rho}{\rho}\right)$$

it is evident that the dominant term in the large r regime (the one of order 1/ρ) is generated by differentiating *only* the trigonometric function at each step. Each such differentiation can be seen to be equivalent to multiplying by (–1) and subtracting p/2 from the argument, so

$$j_l(\rho)\rightarrow\frac{1}{\rho}\sin\left(\rho-\frac{l\pi}{2}\right)\quad \text{as } \rho\rightarrow\infty.$$

These $j_l(\rho)$, then, are the physical partial-wave solutions to the Schrödinger equation with zero potential. When a potential is turned on, the wave function near ‘he origin is still $\sim\rho^l$ (assuming, as we always do, that the potential is negligible compared $l(l+1)/\rho^2$ with the term sufficiently close to the origin). The wave function beyond the range of the potential can be found numerically in principle by integrating out from the origin, and in fact will be like $j_l(\rho)$ above *except* that there will be an extra phase factor, called the "phase shift" and denoted by *d*) in the sine. The significance of this is that in the far region, the wave function is a linear combination of the Bessel function *and* the Neumann function (the solution to the zero-potential Schrödinger equation singular at the origin).

It is therefore necessary to review the Neumann functions as well.

As stated above, the $l = 0$ Neumann function is

$$n_0(\rho) = -\frac{\cos\rho}{\rho},$$

the minus sign being the standard convention.

An argument parallel to the one above for the Bessel functions establishes that the higher-order Neumann functions are given by:

$$n_l(\rho) = (-\rho)^l \left(\frac{1}{\rho}\frac{d}{d\rho}\right)^l \left(-\frac{\cos\rho}{\rho}\right).$$

Near the origin

$$n_l(\rho) \sim \frac{(2l-1)!!}{\rho^{l+1}} \text{ as } \rho \to 0$$

and for large ρ

$$n_l(\rho) \to -\frac{1}{\rho}\cos\left(\rho - \frac{l\pi}{2}\right) \text{ as } \rho \to \infty,$$

so a function of the form $\frac{1}{\rho}\sin\left(\rho - \frac{l\pi}{2} + \delta\right)$ asymptotically can be written as a linear combination of Bessel and Neumann functions in that region.

Finally, the spherical *Hankel* functions are just the combinations of Bessel and Neumann functions that look like outgoing or incoming plane waves in the asymptotic region:

$$h_l(\rho) = j_l(\rho) + in_l(\rho), \quad h_l{}^*(\rho) = j_l(\rho) - in_l(\rho),$$

so for large ρ,

$$h_l(\rho) \to \frac{e^{i(\rho - lx/2)}}{i\rho}, h_l{}^*(\rho) \to -\frac{e^{i(\rho - lx/2)}}{i\rho}$$

THE PARTIAL WAVE SCATTERING MATRIX

Let us imagine for a moment that we could just send in a (time-independent) spherical wave, with q variation given by

$P_l(\cos q)$. For this l th partial wave (dropping overall normalization constants as usual) the radial function far from the origin for zero potential is

$$j_l(kr) \rightarrow \frac{1}{kr}\sin\left(kr - \frac{l\pi}{2}\right) = \frac{i}{2k}\left(\frac{e^{-i(kr-l\pi/2)}}{r} - \frac{e^{+i(kr-l\pi/2)}}{r}\right).$$

If now the (spherically symmetric) potential is turned on, the only possible change to this standing wave solution in the faraway region (where the potential is zero) is a phase shift δ:

$$\sin\left(kr - \frac{l\pi}{2}\right) \rightarrow \sin\left(kr - \frac{l\pi}{2} + \delta_l(k)\right).$$

This is what we would find on integrating the Schrödinger equation out from nonsingular behaviour at the origin.

But in practice, the ingoing wave is given, and *its phase cannot be affected by switching on the potential.* Yet we must still have the solution to the same Schrödinger equation, so to match with the result above we multiply the whole partial wave function by the phase factor $e^{i\delta_l(k)}$. The result is to put *twice* the phase change onto the outgoing wave, so that when the potential is switched on the change in the asymptotic wave function must be

$$\frac{i}{2k}\left(\frac{e^{-i(kr-l\pi/2)}}{r} - \frac{e^{+i(kr-l\pi/2)}}{r}\right) \rightarrow \frac{i}{2k}\left(\frac{e^{-i(kr-l\pi/2)}}{r} - \frac{S_l(k)e^{+i(kr-l\pi/2)}}{r}\right).$$

This equation introduces the *scattering matrix*

$$S_l(k) = e^{2i\delta_l(k)}$$

which must lie on the unit circle $|S| = 1$ to conserve probability the outgoing current must equal the ingoing current. If there is no scattering, that is, zero phase shift, the scattering matrix is unity.

It should be noted that when the radial Schrödinger's equation is solved for a nonzero potential by integrating out from the origin, with $y = 0$ and $y' = 1$ initially, the real function thus generated differs from the wave function given above by an *overall* phase factor $e^{i\delta_l(k)}$.

SCATTERING OF A PLANE WAVE

We're now ready to take the ingoing plane wave, break it into its partial wave components corresponding to different angular momenta, have the partial waves individually phase shifted by *l*-dependent phases, and add it all back together to get the original plane wave plus the scattered wave.

We are only interested here in the wave function far away from the potential. In this region, the original plane wave is

$$e^{i\vec{k}.\vec{r}} = e^{ikr\cos\theta} = \sum_l i^l(2l+1)j_l(kr)P_l(\cos\theta) = \sum_l i^l(2l+1)\frac{i}{2k}\left(\frac{e^{-i(kr-l\pi/2)}}{r} - \frac{e^{+i(kr-l\pi/2)}}{r}\right)P_l(\cos\theta).$$

Switching on the potential phase shifts factor the outgoing wave:

$$\frac{e^{+i(kr-l\pi/2)}}{r} \rightarrow \frac{S_l(k)e^{+i(kr-l\pi/2)}}{r}$$

The actual *scattering* by the potential is the *difference* between these two terms. The complete wave function in the far region (including the incoming plane wave) is therefore:

$$\psi(r,\theta,\varphi) = e^{ikr\cos\theta} + \left(\sum_l (2l+1)\frac{(S_l(k)-1)}{2ik}P_l(\cos\theta)\right)\frac{e^{ikr}}{r}.$$

The i^l factor cancelled the $e^{il\pi/2}$ The -1 $(S_l(k)-1)$ in is there because zero scattering means S = 1. Alternatively, it could be regarded as subtracting off the outgoing waves already present in the plane wave, as discussed above. There is no φ-dependence since with the potential being spherically-symmetric the whole problem is azimuthally-symmetric about the direction of the incoming wave.

It is perhaps worth mentioning that for scattering in just one partial wave, the outgoing current is equal to the ingoing current, whether there is a phase shift or not. So, if switching on the potential does not affect the total current scattered in any partial wave, how can it cause any scattering? The point is that for an ingoing *plane wave* with zero potential, the ingoing and outgoing components have the right relative phase to add to a component of a plane wave a tautology, perhaps. But if an extra phase is introduced into the outgoing wave *only*, the ingoing + outgoing will *no longer give a plane wave* there will be an extra outgoing part proportional to $\left(S_l(k)-1\right)$.

Recall that the scattering amplitude $f(\theta,\varphi)$ was defined in terms of the solution to Schrödinger's equation having an ingoing plane wave by

$$\psi(r,\theta,\varphi)=e^{ikr\cos\theta}+f(\theta,\varphi)\frac{e^{ikr}}{r}.$$

We're now ready to express the scattering amplitude in terms of the partial wave phase shifts for a spherically symmetric potential, of course:

$$f(\theta,\varphi)=f(\theta)=\sum_l(2l+1)\frac{\left(S_l(k)-1\right)}{2ik}P_l(\cos\theta)=\sum_l(2l+1)f_l(k)P_l(\cos\theta)$$

where

$$f_l(k)=\frac{1}{k}e^{i\delta_l(k)}\sin\delta_l(k)$$

is called the *partial wave scattering amplitude,* or just the *partial wave amplitude.*

So the total scattering amplitude is the sum of these partial wave amplitudes:

$$f(\theta)=\frac{1}{k}\sum_l(2l+1)e^{i\delta_l(k)}\sin\delta_l(k)P_l(\cos\theta).$$

The total scattering cross-section

$$\sigma=\int|f(\theta)|^2d\Omega$$

$$=2\pi\int_0^\pi|f(\theta)|^2\sin\theta d\theta$$

$$=2\pi\int_0^\pi\left|\frac{1}{k}\sum_l(2l+1)e^{i\delta_l(k)}\sin\delta_l(k)P_l(\cos\theta)\right|^2\sin\theta d\theta$$

and the normalization of the Legendre polynomials

$$\int_0^\pi P_l^2(\cos\theta)\sin\theta d\theta = \frac{2}{2l+1}$$

gives

$$\sigma = 4\pi\sum_{l=0}^{\infty}(2l+1)|f_l(k)|^2 = \frac{4\pi}{k^2}\sum_{l=0}^{\infty}(2l+1)\sin^2\delta_l.$$

So the total cross-section is the sum of the cross-sections for each l value. This does *not* mean, though, that the differential cross-section for scattering *into a given solid angle* is a sum over separate l values the different components interfere. It is only when *all* angles are integrated over that the orthogonality of the Legendre polynomials guarantees that the cross-terms vanish.

Notice that the maximum possible scattering cross-section for particles in angular momentum state l is $(4\pi/k^2)(2l+1)$ which is *four times* the classical cross section for that partial wave impinging on, say, a hard sphere! (Imagine semiclassically particles in an annular area: angular momentum $L = rp$, say, but $L = \hbar l$ and $p = \hbar k$ so $l = rk$. Therefore the annular area corresponding to angular momentum "between" l and $l + 1$ has inner and outer radii l/k and $(l+1)/k$ and therefore area $2\pi(2l+1)/k^2$.) The quantum result is essentially a diffractive effect, we'll discuss it more later.

It's easy to prove the *optical theorem* for a spherically-symmetric potential: just take the imaginary part of each side of the equation

$$f(\theta) = \frac{1}{k}\sum_l (2l+1)e^{i\delta_l(k)}\sin\delta_l(k)P_l(\cos\theta)$$

at q = 0, using $P_l(1) = 1$,

$$\mathrm{Im}\, f(\theta = 0) = \frac{1}{k}\sum_l (2l+1)\sin^2\delta_l(k)$$

from which the optical theorem $\mathrm{Im}\, f(0) = k\sigma/4\pi$ follows immediately.

It's also worth noting what the unitarity of the lth partial wave scattering matrix $S_l^{\dagger} S_l = 1$ implies for the partial wave amplitude $f_l(k) = \frac{1}{k} e^{i\delta_l(k)} \sin\delta_l(k)$. Since $\delta_l(k) = e^{2i\delta_l(k)}$, it follows that

$$S_l(k) = 1 + 2ikf_l(k).$$

From this, $S_l^{\dagger} S_l = 1$ gives:

$$\text{Im}\, f_l(k) = k\left|f_l(k)\right|^2.$$

This can be put more simply:

$$\text{Im}\,\frac{1}{f_l(k)} = -k.$$

In fact,

$$f_l(k) = \frac{1}{k(\cot\delta_l(k) - i)}.$$

PHASE SHIFTS AND POTENTIALS: SOME EXAMPLES

We assume in this section that the potential can be taken to be zero beyond some boundary radius *b*. This is an adequate approximation for all potentials found in practice *except* the Coulomb potential, which will be discussed separately later.

Asymptotically, then,

$$\psi_l(r) = \frac{i}{2k}\left(\frac{e^{-i(kr - l\pi/2)}}{r} - \frac{e^{2i\delta_l(k)} e^{+i(kr - l\pi/2)}}{r}\right)$$

$$= \frac{e^{i\delta_l(k)}}{kr}\sin\left(kr + \delta_l(k) - l\pi/2\right)$$

$$= \frac{e^{i\delta_l(k)}}{kr}\left(\sin(kr - l\pi/2)\cos\delta_l(k) + \cos(kr - l\pi/2)\sin\delta_l(k)\right).$$

This expression is only exact in the limit $r \to \infty$ but since the potential can be taken zero beyond $r = b$, the wave function must have the form

$$\psi_l(r) = e^{i\delta_l(k)}\left(\cos\delta_l(k)\, j_l(kr) - \sin\delta_l(k)\, n_l(kr)\right)$$

for $r > b$.

THE HARD SPHERE

The simplest example of a scattering potential:

$$V(r) = \infty \text{ for } r < R,$$

$$V(r) = 0 \text{ for } r \geq R.$$

The wave function must equal zero at $r = R$, so from the above form of $\psi_l(r)$,

$$\tan\delta_l(k) = \frac{j_l(kR)}{n_l(kR)}$$

For $l = 0$, $\tan\delta_0(k) = -\frac{(\sin kR)/kR}{(\cos kR)/kR} = -\tan kR$, so $\delta_0(k) = -kR$. This amounts to the wave function being effectively moved over to begin at R instead of at the origin:

$$\frac{\sin kr}{kr} \to \frac{\sin k(r+\delta)}{kr} = \frac{\sin k(r-R)}{kr}$$

for $r > R$, of course y = 0 for $r < R$.

For higher angular momentum states at low energies ($kR \ll 1$),

$$\tan\delta_l = \frac{j_l(kR)}{n_l(kR)} \approx -\frac{(kR)^l/(2l+1)!!}{(2l-1)!!/(kR)^{l+1}} = -\frac{(kR)^{2l+1}}{(2l+1)((2l-1)!!)^2}.$$

Therefore at low enough energy, only $l = 0$ scattering is important as is obvious, since an incoming particle with momentum $p = \hbar k$ and angular momentum $l\hbar$ is most likely at a distance l/k from the center of the potential at closest approach, so if this is much greater than R, the phase shift will be small.

THE BORN APPROXIMATION FOR PARTIAL WAVES

From the definition of $f(\theta,\varphi)$

$$\psi_k(r) = e^{ikr} + f(\theta,\varphi)\frac{e^{ikr}}{r}$$

and

$$\psi_{\mathbf{k}}(\mathbf{r}) = e^{i\mathbf{k}\mathbf{r}} - \frac{m}{2\pi\hbar^2}\frac{e^{ikr}}{r}\int d^3r' e^{-i\mathbf{k}_f \mathbf{r}'}V(\mathbf{r}')\psi_{\mathbf{k}}(\mathbf{r}')$$

recall the Born approximation amounts to replacing the wave function $\psi_k(r')$ in the integral on the right by an outgoing plane wave, therefore ignoring rescattering, or the distortion of the outgoing wave by the potential.

To translate this into a partial wave approximation, we first express take the incoming k to be in the z-direction, so in the integrand

$$\psi_{\mathbf{k}}(\mathbf{r}') = e^{ikr'\cos\theta} = \sum_l i^l(2l+1)j_l(kr')P_l(\cos\theta').$$

Labeling the angle between k_f and r′ by γ,

$$e^{-i\mathbf{k}_f\mathbf{r}'} = \sum_l (-i)^l(2l+1)j_l(kr')P_l(\cos\gamma).$$

Now k_f is in the direction (θ,φ) and r′ in the direction (θ',φ'), and g is the angle between them. For this situation, there is an *addition theorem for spherical harmonics*:

$$P_l(\cos\gamma) = \frac{4\pi}{2l+1}\sum_{m=-l}^{l} Y^*_{lm}(\theta',\varphi')Y_{lm}(\theta,\varphi).$$

On inserting this expression and integrating over θ',φ', the nonzero m terms give zero, in fact the only nonzero term is that with the same l as the term in the $\psi_k(r')$ expansion, giving

$$f(\theta) = -\frac{2m}{\hbar^2}\sum_{l=0}^{\infty}(2l+1)P_l(\cos\theta)\int_0^{\infty} r^2 dr V(r)\left(j_l(kr)\right)^2$$

and remembering

$$f(\theta) = \frac{1}{k}\sum_l (2l+1)e^{i\delta_l(k)}\sin\delta_l(k)P_l(\cos\theta)$$

it follows that for small phase shifts (the only place it's valid) the partial-wave Born approximation reads

$$\delta_l \approx -\frac{2mk}{\hbar^2}\int_0^\infty r^2 dr V(r)\big(j_l(kr)\big)^2$$

LOW ENERGY SCATTERING: THE SCATTERING LENGTH

From $f_l(k) = \dfrac{1}{k\left(\cot\delta_l(k) - i\right)}$, the $l = 0$ cross section is

$$\sigma_{l=0} = \frac{4\pi}{k^2\left|\cot\delta_0 - i\right|^2}.$$

At energy $E \to 0$ the radial Schrödinger equation for $u = r\psi$ away from the potential becomes $d^2u/dr^2 = 0$, with a straight line solution $u(r) = C(r-a)$. This must be the $k \to 0$ limit of $u(r) = C'\sin(kr + \delta_0)$, which can only be correct if δ_0 is itself linear in k for sufficiently small k, and then it must be $\delta_0 = -ka$ a being the point at which the extrapolated external wavefunction intersects the axis (maybe at negative r!) So, as k goes to zero, the cot term dominates in the denominator and

$$\sigma_{l=0}(k \to 0) = 4\pi a^2$$

The quantity a is called the *scattering length.*

Integrating the zero-energy radial Schrödinger equation out from $u(r) = 0$ at the origin for a weak (spherical) square well potential, it is easy to check that a is positive for a repulsive potential, negative for an attractive potential.

Scattering Length for Square Well

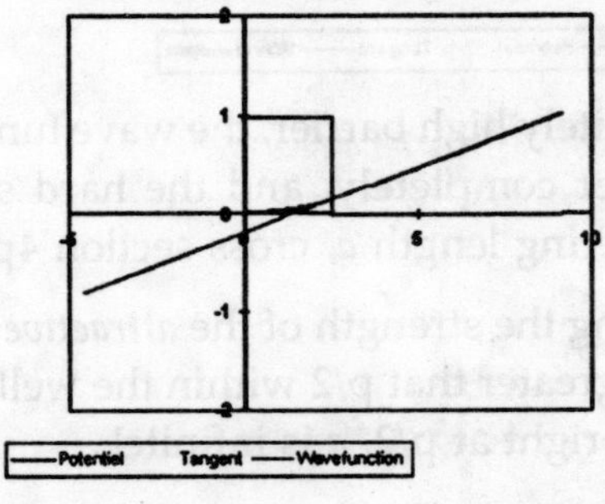

Potential Tangent Wavefunction

Repulsive potential, zero-energy wave function (so it's a *straight line* outside of the well!):

Attractive potential:

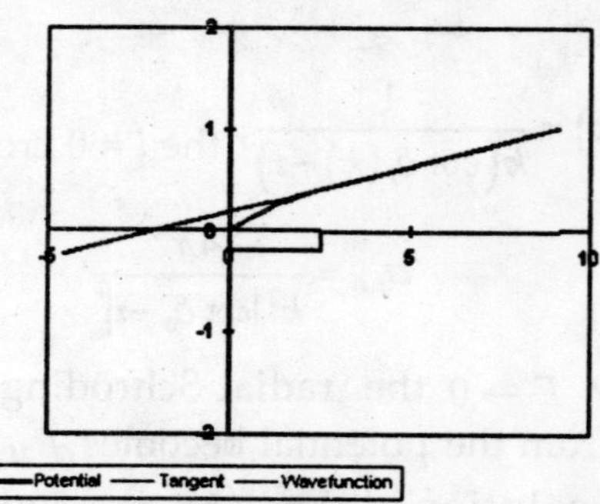

On increasing the strength of the repulsive potential, still solving for the zero-energy wave function, *a* tends to the potential wall here's the zero-energy wavefunction for a barrier of height 6:

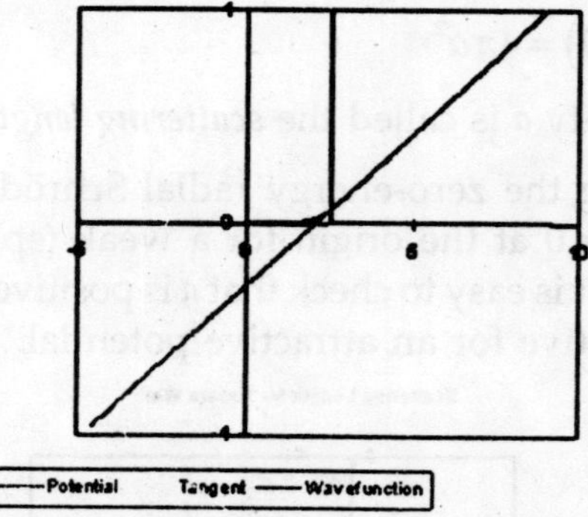

For an infinitely high barrier, the wave function is pushed out of the barrier completely, and the hard sphere result is recovered: scattering length *a*, cross-section $4pa^2$.

On increasing the strength of the *attractive* well, if there is a phase change greater that p/2 within the well, *a* will become positive. In fact, right at p/2, *a* is infinite!

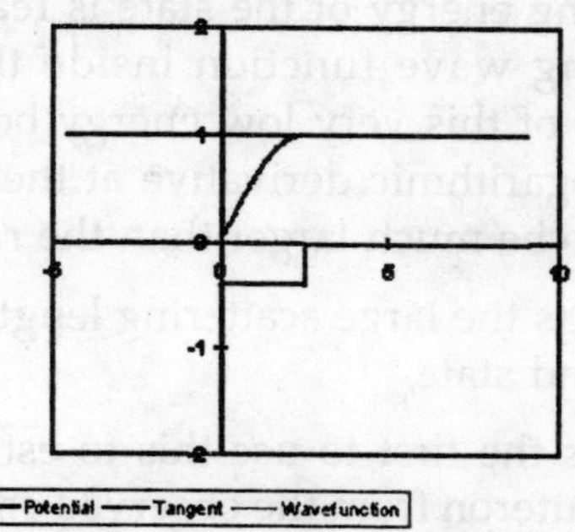

And a little more depth to the well gives a *positive* scattering length:

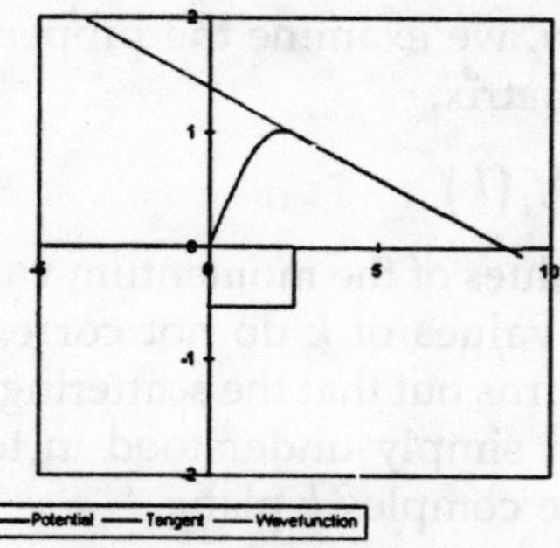

In fact, a well deep enough to have a positive scattering length will also have a bound state. This becomes evident when one considers that the depth at which the scattering length becomes infinite can be thought of as formally having a zero energy bound state, in that although the wave function outside is not normalizable, it is equivalent to an exponentially decaying function with infinite decay length. If one now deepens the well a little, the zero-energy wave function inside the well curves a little more rapidly, so the slope of the wave function at the edge of the well becomes negative, as in the picture above. With this slightly deeper well, we can now *lower* the energy slightly to negative values. This will have little effect on the wave function inside the well, but make possible a fit at the well edge to an exponential decay outside a genuine bound state, with wave function outside the well.

If the binding energy of the state is really low, the zero-energy scattering wave function inside the well is almost identical to that of this very low energy bound state, and in particular the logarithmic derivative at the wall will be very close, taking a to be much larger than the radius of the well.

This connects the large scattering length to the energy of the weakly bound state,

Wigner was the first to use this to estimate the binding energy of the deuteron from the observed cross section for low energy neutron-proton scattering.

LOW ENERGY APPROXIMATIONS FOR THE S MATRIX

In this section, we examine the properties of the partial-wave scattering matrix.

$$S_l(k) = 1 + 2ikf_l(k)$$

for *complex* values of the momentum variable k. Of course, general complex values of k do not correspond to physical scattering, but it turns out that the scattering of physical waves can often be most simply understood in terms of dominant singularities in the complex k plane.

We begin with the complex k connection between (positive energy) scattering and (negative energy) bound states. The asymptotic form of the $l = 0$ partial wave function in a scattering experiment is

$$\frac{e^{-i(kr-l\pi/2)}}{r} - \frac{S_0(k)e^{+i(kr-l\pi/2)}}{r}$$

An $l = 0$ bound state has asymptotic wave function

$$\frac{e^{-xr}}{r}$$

Notice that this resembles an "outgoing wave" with imaginary momentum $k = ik$. If we analytically continue the *scattering* wave function from real k into the complex k-plane, we get *both* exponentially increasing and decreasing wave functions, making no physical sense. But there is an exception to this general observation: if the scattering matrix $S_0(k)$ becomes infinite at some complex value of k, the exponentially decreasing term will dominate, and we'll only have a decreasing wave function a bound state. We know that the energy of a bound state has to be real and negative, and also equal to $\hbar^2\hbar^2/2m$, so this can only happen for k *pure imaginary, k = ik.*

Now, the existence of a low energy bound state means that the S matrix has a pole (on the imaginary axis) close to the origin, so this will strongly affect low energy (near the origin, but real k) scattering. Let's see how that works using the low-energy approximation discussed previously. Recall that the $l = 0$ partial wave amplitude

$$f_0(k) = \frac{1}{k\{\cot\delta_0(k) - i\}},$$

and at low energy $\delta_0(k) = -ka$ so

$$f_0(k) = \frac{1}{k((-1/ka) - i)} = -\frac{1}{ik + 1/a},$$

and

$$S_0(k) = 1 + 2ikf_0(k) = -\frac{k + (i/a)}{k - (i/a)}$$

Note that $S \to 1$ as $k \to 0$, as it should, since $\delta_0(k) = -ka \to 0$ and $S_0(k) = e^{2i\delta_0(k)}$. Note also that this approximation correctly gives $|S_0(k)| = 1$.

This $S_0(k)$ has a pole in the complex plane at $k = i/a$, and if this corresponds to a bound state having $k = 1/a$, then the binding energy $\hbar^2k^2/2m = \hbar^2/2ma^2$. In fact, though, we run into a problem here: we get the same form of $S_0(k)$ at low energies even for a *repulsive* potential, which certainly doesn't have a bound state! The pole in $S_0(k)$ only means that we can have an asymptotic wave function of the right form, but it does *not* guarantee that this asymptotic form will go smoothly to nonsingular behaviour at the origin. For a repulsive potential, it's easy to see that the zero-energy wave function on integrating out from the origin slopes more and more steeply *upwards,* so could never, with increasing *r*, go over to asymptotic decay.

EFFECTIVE RANGE

The low energy approximation above can be written $k\cot\delta_0(k) = -1/a$. We shall now derive a better approximation, $k\cot\delta_0(k) = -(1/a) + \frac{1}{2}r_0k^2$, where r_0, called the "effective range", gives some measure of the extent of the potential (in contrast to *a,* which, as we have seen can be arbitrarily large, even for a short-range potential).

A useful mathematical tool needed at this point is the *Wronskian*. For two functions $f(x), g(x)$ the Wronskian is defined as $W(f,g) = fg' - f'g$, the prime denoting differentiation as usual. From this, $W'(f,g) = fg'' - f''g$, and if $f(x), g(x)$ satisfy the same second-order differential equation (like the Schrödinger equation with the same energy) then $W(f,g)$ is a constant, independent of *x*.

For the radial Schrödinger equation, asymptotically

$$u(k,r) \to C\sin(kr + \delta_0) \equiv v(k,r)$$

where we now show *k* explicitly. This asymptotic function *v*(*k, r*) satisfies the Schrödinger equation for *zero potential,* but does *not* have the correct physical boundary behaviour at *r* = 0.

Since in the low-energy limit $\delta_0(k) = -ka$ the $k = 0$ asymptotic wave function $v(0,r) = 1 - r/a$ (taking $C = 1/\sin\delta_0$).

From the Schrödinger equation

$$-u''(k,r)+\left(2mV(r)/\hbar^2\right)u(k,r)=k^2u(k,r)$$

it is easy to check that the Wronskian $u(k,r)$ of with the corresponding *zero energy* function $u(0,r)$ satisfies:

$$\frac{d}{dr}W\left[u(k,r),u(0,r)\right]=k^2u(k,r)u(0,r).$$

(The term involving the potential has canceled out.)

The corresponding functions $v(k,r),v(0,r)$ satisfy the same Wronskian equation:

$$\frac{d}{dr}W\left[v(k,r),v(0,r)\right]=k^2v(k,r)v(0,r).$$

We can find a formula for the effective range r_0 by integrating the difference between these two equations from $r = 0$ to infinity: the two solutions *u*, *v* differ only within the range of the potential, and appropriately normalizing them, then taking the difference, gives a measure of this range. So,

$$\left\{W\left[v(k,r),v(0,r)\right]-W\left[u(k,r),u(0,r)\right]\right\}_{r=0}^{r=\infty}=k^2\int_0^\infty\left[v(k,r)v(0,r)-u(k,r)u(0,r)\right]dr$$

For r large, $u(k,r)\to v(k,r)$, so there is zero contribution from the upper end. For, the properly-behaved *u* functions go to zero, the *v* functions are

$$v(k,r)=C\sin(kr+\delta_0)=\sin(kr+\delta_0)/\sin\delta_0$$

from which, with $\delta_0(k)=-ka, v(0,r)=1-r/a$. It follows immediately that

$$W\left[v(k,r),v(0,r)\right]_{r=0}=-\frac{1}{a}-k\cot\delta_0$$

and therefore

$$k\cot\delta_0=-\frac{1}{a}+k^2\int_0^\infty\left[v(k,r)v(0,r)-u(k,r)u(0,r)\right]dr$$

with a low-energy limit

$$k\cot\delta_0 = -\frac{1}{a}+\frac{1}{2}k^2 r_0$$

where

$$r_0 = 2\int_0^\infty \left[v^2(0,r)-u^2(0,r)\right]dr\,.$$

Now, by definition $u(0,r), v(0,r)$ coincide outside the range of the potential, but moving from that region towards the origin, they part company when the potential kicks in, with $u\to 0, v\to 1$ as $r\to 0$. Therefore the integral above is a rough measure of the actual *range* of the potential about half of it (hence the factor of 2 in defining r_0). Note again the contrast with *a*, which can be infinite for a short range potential.

COULOMB SCATTERING AND THE HYDROGEN ATOM BOUND STATES

One particular set of bound states in a potential we've spent a good deal of time on are the states of the hydrogen atom, and it is interesting to see how they relate to scattering. Recall that the asymptotic form of the bound state wave function is:

$$R_{nl(r)} \sim r^n \frac{e^{-r/na_0}}{r}$$

But this doesn't have the bound-state form we found above from the analytic continuation argument, there's an extra r^n! What's going on? The problem is that in all our previous work, we assumed that if we looked far enough away from the center of the potential, the radial Schrödinger equation could be taken to be that for zero potential, to any desired accuracy. The Coulomb potential, though, does not decay fast enough with distance for this to be true. For instance, it has bound states having arbitrarily large radii. Writing

$$\kappa = \frac{1}{na_0} = \frac{me^2}{n\hbar^2}$$

we can write

$$R_{nl}(r) \sim \frac{1}{r} e^{-kr + \left(me^2 / \hbar^2 k\right) \ln r}$$

Note that the extra term in the exponent keeps on growing, without limit! We are never free of the potential.

But how does this analysis of hydrogen atom wave functions relate to positive-energy scattering states? We can just analytically continue this result back to real *k* to find out. Replacing *-k* by *ik* gives:

$$R_{nl}(r) \sim \frac{1}{r} e^{i\left(kr + \left(me^2 / \hbar^2 k\right) \ln r\right)}.$$

So we have scattering states that are not of the standard form either the phase shift is infinite, and not well-defined. But we found this result be analytically continuing from the hydrogen atom bound states. Let's check it: let us look at the radial Schrödinger equation for positive energies at large *r*. Writing $R(r) = u(r)/r$ as usual, let us also put $u(r) = e^{ikr}v(r)$ for large *r*, and we can also ignore the centrifugal barrier term, so

$$-\frac{\hbar^2}{2m}u'' - \frac{e^2}{r}u = Eu = \frac{\hbar^2 k^2}{2m}u.$$

The equation for $v(r)$ is

$$-\frac{\hbar^2}{2m}\left(2ikv' + v''\right) - \frac{e^2}{r}v = 0$$

and since $v(r)$ is slowly varying, the second derivative can be ignored, so

$$v' \cong \frac{ime^2}{\hbar^2 kr} v$$

This leads immediately to the same form we found by analytic continuation.

RESONANCES AND ASSOCIATED ZEROS

Recall in the first semester we discussed a-decay: an a particle in a heavy nucleus can be thought of as trapped in a potential well generated by the attractive nuclear forces. A spherical square well is a workable approximation, except that this square well is at the top of a hill outside the nucleus, the repulsive electrostatic potential is $(Z-2)2e^2/r$, sloping down from the well edge to zero as $r\to\infty$. This means that fro a radioactive nucleus, although the energy level would be at negative energy for the square well on level ground, actually it's above the bottom of the electrostatic hill, and the $r\to\infty$ wave function will not be decaying but oscillating. This asymptotic wave is of course very tiny, since typically the chances of detecting the a well outside the nucleus, that is, of decay, is one in millions of years.

Now consider the reverse process: imagine we bombard a decayed nucleus with a particles. If we sent in one at exactly the right energy (very difficult this is a thought experiment!) the wave function would be exactly the same as that for a decay. The wave function inside the nucleus would be huge compared with that outside, we'd never see our a again. Less dramatically, if we sent in one close to that energy, the wave function would still be very large inside the nucleus, meaning that the particle would spend a long time inside before coming out again. (Recall for a particle in a roller coaster potential in one dimension, the wave function is large where the particle spends a lot of time that's where you're most likely to find it.) This is a *resonance*: at just the right energy, the amplitude of the wave function within the potential becomes very large, analogous to the amplitude of a classical driven oscillator as the driving frequency is adjusted to the natural oscillator frequency.

Can we understand this in terms of poles in the *S* matrix? Considered as a function of energy, the *S* matrix has poles at negative energies corresponding to bound states. But this is a positive energy and $|S(k)|=1$, so it can't have a pole at a positive energy. What it can have is a pole *near* a positive energy, in the complex plane. To keep $|S(k)|=1$, it would then have to have a zero at the mirror image point, that is, be locally of the form.

$$S_l(E) = e^{2i\delta_l(E)} = \frac{E - E_0 - i\Gamma/2}{E - E_0 + i\Gamma/2}$$

From this, $\tan\delta_1(E) = -\Gamma/2(E - E_0)$, so $\delta_l(E_0) = \pi/2$, and the scattering cross section reaches its maximum possible value, recall $\sigma = 4\pi\sum_{l=0}^{\infty}(2l+1)|f_l(k)|^2 = \frac{4\pi}{k^2}\sum_{l=0}^{\infty}(2l+1)\sin^2\delta_l$, so

$$\sigma_l^{max} = \frac{4\pi}{k^2}(2l+1).$$

For a narrow resonance (small G) the phase shift $\delta_1(E)$ goes rapidly from 0 to p as the energy is increased through E_0. Most of the variation occurs within an energy range Γ of E_0, Γ is called the *width* of the resonance. If the resonance is superimposed on a slowly varying background phase shift d, then it causes an increase from d to δ + p. This will pass through 0 or p, depending on the initial sign of d, so the maximum scattering at phase shift p/2 will have associated with it an energy at which there is zero scattering. For substantial background δ, the zero could be close to the peak, as illustrated below:

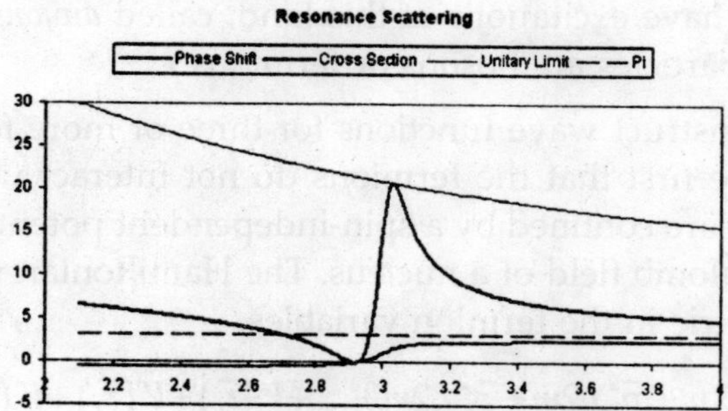

INTRODUCTION

For two identical particles confined to a one-dimensional box, we established earlier that the normalized two-particle wavefunction $\psi(x_1, x_2)$, which gives the probability of finding simultaneously one particle in an infinitesimal length dx_1 at

x_1 and another in dx_2 at x_2 as $|\psi(x_1,x_2)|^2 dx_1dx_2$, only makes sense if $|\psi(x_1,x_2)|^2 = |\psi(x_2,x_1)|^2$, since we don't know which of the two indistinguishable particles we are finding where.

It follows from this that there are two possible wave function symmetries: $\psi(x_1,x_2)=\psi(x_2,x_1)$ or $\psi(x_1,x_2)=-\psi(x_2,x_1)$. It turns out that if two identical particles have a symmetric wave function in some state, particles of that type always have symmetric wave functions, and are called *bosons*. (If in some other state they had an antisymmetric wave function, then a linear superposition of those states would be neither symmetric nor antisymmetric, and so could not satisfy $|\psi(x_1,x_2)|^2 = \left|\psi(x_2,x_1)^2\right|$.) Similarly, particles having antisymmetric wave functions are called *fermions*. (Actually, we could in principle have $\psi(x_1,x_2)=e^{i\alpha}\psi(x_2,x_1)$, with a a constant phase, but then we wouldn't get back to the original wave function on exchanging the particles twice. Some two-dimensional theories used to describe the quantum Hall effect do in fact have excitations of this kind, called *anyons*, but all ordinary particles are bosons or fermions.)

To construct wave functions for three or more fermions, we assume first that the fermions do not interact with each other, and are confined by a spin-independent potential, such as the Coulomb field of a nucleus. The Hamiltonian will then be symmetric in the fermion variables,

$$H = \vec{p}_1^{\,2}/2m + \vec{p}_2^{\,2}/2m + \vec{p}_3^{\,2}/2m + \ldots + V(\vec{r}_1) + V(\vec{r}_2) + V(\vec{r}_3) + \ldots$$

and the solutions of the Schrödinger equation are products of eigenfunctions of the single-particle Hamiltonian $H = \vec{p}^{\,2}/2m + V(\vec{r})$. However, these products, for example $\psi_a(1)\psi_b(2)\psi_c(3)$ do not have the required antisymmetry property. Here *a, b, c, …* label the single-particle eigenstates, and 1, 2, 3, … denote both space and spin coordinates of single particles, so 1 stands for $(\vec{r}_1, s_1)$. The necessary antisymmetrization for the particles 1, 2 is achieved by subtracting the same product wave function with the particles 1 and 2 interchanged, $\psi_a(1)\psi_b(2)\psi_c(3)$ so is replaced by $\psi_a(1)\psi_b(2)\psi_c(3) - \psi_a(2)\psi_b(1)\psi_c(3)$, ignoring overall normalization for now.

But of course the wave function needs to be antisymmetrized with respect to *all* possible particle exchanges, so for 3 particles we must add together all 3! permutations of 1, 2, 3 in the state *a, b, c,* with a factor -1 for each particle exchange necessary to get to a particular ordering from the original ordering of 1 in *a*, 2 in *b*, and 3 in *c*. In fact, such a sum over permutations is precisely the *definition* of the determinant, so, with the appropriate normalization factor:

$$\psi_{abc}(1,2,3) = \frac{1}{\sqrt{3!}} \begin{vmatrix} \psi_a(1) & \psi_b(1) & \psi_c(1) \\ \psi_a(2) & \psi_b(2) & \psi_c(2) \\ \psi_a(3) & \psi_b(3) & \psi_c(3) \end{vmatrix}$$

where *a, b, c* label three (different) quantum states and 1, 2, 3 label the three fermions. The determinantal form makes clear the antisymmetry of the wave function with respect to exchanging any two of the particles, since exchanging two rows of a determinant multiplies it by -1.

We also see from the determinantal form that the three states *a, b, c* must all be different, for otherwise two columns would be identical, and the determinant would be zero. This is just Pauli's Exclusion Principle: no two fermions can be in the same state. Although these determinantal wave functions are only strictly correct for noninteracting fermions, they are a useful beginning in describing electrons in atoms (or in a metal), with the electron-electron repulsion approximated by a single-particle potential. For example, the Coulomb field in an atom, as seen by the outer electrons, is partially shielded by the inner electrons, and a suitable *V*(*r*) can be constructed self-consistently, by computing the single-particle eigenstates and finding their associated charge densities.

SPACE AND SPIN WAVE FUNCTIONS

Suppose we have two electrons in some spin-independent potential $V(r)$ (for example in an atom). We know the two-electron wave function is antisymmetric. Now, the Hamiltonian has no spin-dependence, so we must be able to construct a set of common eigenstates of the Hamiltonian, the total spin, and the z-component of the total spin.

$$\chi_S(s_1, s_2) = |S_{tot} = 0, S_z = 0\rangle = \left(1/\sqrt{2}\right)\left(|\uparrow\downarrow\rangle - |\downarrow\uparrow\rangle\right)$$

For two electrons, there are four basis states in the spin space. The eigenstates of S and S_z are the singlet state

$$\chi_T^1(s_1, s_2) = |1,1\rangle = |\uparrow\uparrow\rangle, \quad |1,0\rangle = \left(1/\sqrt{2}\right)\left(|\uparrow\downarrow\rangle + |\downarrow\uparrow\rangle\right), \quad |1,-1\rangle = |\downarrow\downarrow\rangle$$

and the triplet states

where the first arrow in the ket refers to the spin of particle 1, the second to particle 2.

$$\Psi(\vec{r}_1, \vec{r}_2, s_1, s_2) = \psi(\vec{r}_1, \vec{r}_2)\, \chi(s_1, s_2)$$

It is evident by inspection that the singlet spin wave function is antisymmetric in the two particles, the triplet symmetric. The total wave function for the two electrons in a common eigenstate of S, S_z and the Hamiltonian H has the form:

and Y must be antisymmetric. It follows that a pair of electrons in the *singlet* spin state must have a *symmetric spatial wave function,,* whereas electrons in the *triplet* state, that is, with their spins parallel, have an *antisymmetric* spatial wave function.

DYNAMICAL CONSEQUENCES OF SYMMETRY

This overall antisymmetry requirement actually determines the magnetic properties of atoms. The electron's magnetic moment is aligned with its spin, and *even though the spin variables do not appear in the Hamiltonian, the energy of the eigenstates depends on the relative spin orientation*. This arises from the electrostatic repulsion energy between the electrons. In the spatially antisymmetric state, the two electrons have zero probability of being at the same place, and are on average further apart than in the spatially symmetric state. Therefore, the electrostatic repulsion raises the energy of the spatially symmetric state above that of the spatially antisymmetric state. It follows that *the lower energy state has the spins pointing in the same direction*. This argument is still valid for more than two electrons, and leads to Hund's rule for the magnetization of incompletely filled inner shells of electrons in transition metal atoms and rare earths: if the shell is half filled or less, all the spins point in the same direction. This is the first step in understanding ferromagnetism.

Another example of the importance of overall wave function antisymmetry for fermions is provided by the specific heat of hydrogen gas. This turns out to be heavily dependent on whether the two protons (spin one-half) in the H_2 molecule have their spins parallel or antiparallel, even though that alignment involves only a very tiny interaction energy. If the proton spins are antiparallel, that is to say in the singlet state, the molecule is called parahydrogen. The triplet state is called orthohydrogen. These two distinct gases are remarkably stable in the absence of magnetic impurities, para–ortho transitions take weeks.

The actual energy of interaction of the proton spins is of course completely negligible in the specific heat. The important contributions to the specific heat are the usual kinetic energy term, and the rotational energy of the molecule. This is where the overall (space‰spin) antisymmetric wave function for the protons plays a role. Recall that the parity of a state with rotational angular momentum l is $(-1)^l$. Therefore, parahydrogen, with an antisymmetric proton *spin* wave function, must have a symmetric proton *space* wave function, and so can only have *even* values of the rotational angular momentum. Orthohydrogen can only have odd values. The energy of the rotational level with angular momentum l is $E_l^{rot} = \hbar^2 l(l+1)/I$, so the two kinds of hydrogen gas have different sets of rotational energy levels, and consequently different specific heats.

SYMMETRY OF THREE-ELECTRON WAVE FUNCTIONS

Things get trickier when we go to three electrons. There are now $2^3 = 8$ basis states in the spin space. Four of these are accounted for by the spin 3/2 state with all spins pointing in the same direction. This is evidently a symmetric state, so must be multiplied by an antisymmetric spatial wave function, a determinant. But the other four states are two pairs of total spin ½ states. They are orthogonal to the symmetric spin 3/2 state, so they can't be symmetric, but they can't be antisymmetric either, since in each such state two of the spins must be pointing in the same direction! An example of such a state is

$$\chi(s_1, s_2, s_3) = |\uparrow_1\rangle(1/\sqrt{2})(|\uparrow_2\downarrow_3\rangle - |\downarrow_2\uparrow_3\rangle).$$

Evidently, this must be multiplied by a spatial wave function symmetric in 2 and 3, but to get a total wave function with *overall* antisymmetry it is necessary to add more terms:

$$\Psi(1,2,3) = \chi(s_1,s_2,s_3)\psi(\vec{r}_1,\vec{r}_2,\vec{r}_3) + \chi(s_2,s_3,s_1)\psi(\vec{r}_2,\vec{r}_3,\vec{r}_1) + \chi(s_3,s_1,s_2)\psi(\vec{r}_3,\vec{r}_1,\vec{r}_2)$$

(from Baym). Requiring the spatial wave function $\psi(\vec{r}_1,\vec{r}_2,\vec{r}_3)$ to be symmetric in 2, 3 is sufficient to guarantee the overall antisymmetry of the total wave function Ψ. Particle enthusiasts might be interested to note that functions exactly like this arise in constructing the spin/flavour wave function for the proton in the quark model.

For more than three electrons, similar considerations hold. The mixed symmetries of the spatial wave functions and the spin wave functions which together make a totally antisymmetric wave function are quite complex, and are described by Young diagrams (or tableaux).

SCATTERING OF IDENTICAL PARTICLES

As a preliminary exercise, consider the *classical* picture of scattering between two positively charged particles, for example a-particles, viewed in the center of mass frame. If an outgoing a is detected at an angle q to the path of ingoing a

#1, it could be #1 deflected through q, or #2 deflected through p - q . Classically, we could tell which one it was by watching the collision as it happened, and keeping track.

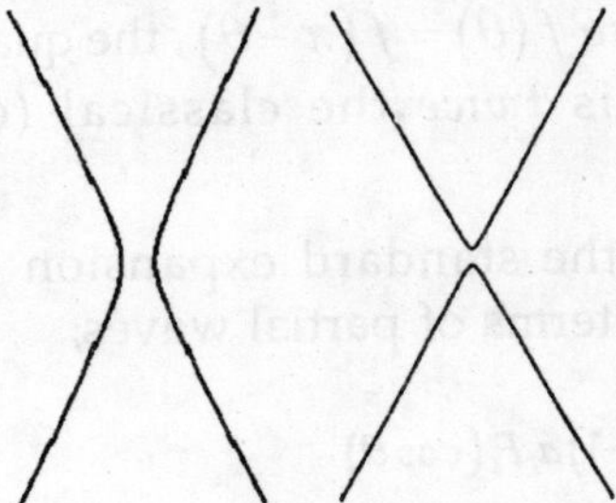

However, in a quantum mechanical scattering process, we cannot keep track of the particles unless we bombard them with photons having wavelength substantially less than the distance of closest approach. This is just like detecting an electron at a particular place when there are two electrons in a one dimensional box: the probability *amplitude* for finding an a coming out at angle q to the ingoing direction of one of them is the sum of the amplitudes (*not* the sum of the probabilities!) for scattering through q and p - q.

Writing the asymptotic scattering wave function in the standard form for scattering from a fixed target,

$$\psi(\vec{r}) \approx e^{ikz} + f(\theta)\frac{e^{ikr}}{r}$$

the two-particle wave function in the center of mass frame, in terms of the relative coordinate, is given by symmetrizing:

$$\psi(\vec{r}) \approx e^{ikz} + e^{-ikz} + \left(f(\theta) + f(\pi - \theta)\right)\frac{e^{ikr}}{r}.$$

How does the particle symmetry affect the actual scattering rate at an angle q? If the particles were distinguishable, the differential cross section would be

$$\left(\frac{d\sigma}{d\Omega}\right)_{\text{distinguishable}} = |f(\theta)|^2 + |f(\pi - \theta)|^2$$

but quantum mechanically

$$\left(\frac{d\sigma}{d\Omega}\right) = \left|f(\theta) + f(\pi - \theta)\right|^2.$$

This makes a big difference! For example, for scattering through 90°, where $f(\theta) = f(\pi - \theta)$, the quantum mechanical scattering rate is *twice* the classical (distinguishable) prediction.

if we make the standard expansion of the scattering amplitude f(q) in terms of partial waves,

$$f(\theta) = \sum_{l=0}^{\infty} (2l+1) a_l P_l(\cos\theta)$$

then

$$f(\theta) + f(\pi - \theta) = \sum_{l=0}^{\infty} (2l+1) a_l \left(P_l(\cos\theta) + P_l(\cos(\pi - \theta))\right)$$

$$= \sum_{l=0}^{\infty} (2l+1) a_l \left(P_l(\cos\theta) + P_l(-\cos\theta)\right)$$

and since $p_l(-x) = (-1)^l P_l(x)$ the scattering only takes place in even partial wave states. This is the same thing as saying that the overall wave function of two identical bosons is symmetric, so if they are in an eigenstates of total angular momentum, from $p_l(-x) = (-1)^l P_l(x)$ it has to be a state of even *l*.

For fermions in an antisymmetric spin state, such as proton-proton scattering with the two proton spins forming a singlet, the spatial wave function is symmetric, and the argument is the same as for the boson case above. For *parallel* spin protons, however, the spatial wave function has to be antisymmetric, and the scattering amplitude will then be $f(\theta) - f(\pi - \theta)$ In this case there is *zero* scattering at 90°!

Note that for (nonrelativistic) equal mass particles, the scattering angle in the center of mass frame is twice the scattering angle in the fixed target (lab) frame. This is easily seen in the diagram below. The four equal-length black arrows, two in, two out, forming an X, are the center of mass momenta. The lab momenta are given by adding the (same length) blue dotted arrow to each, reducing one of the ingoing momenta to zero, and giving the (red arrow) lab momenta. The outgoing lab momenta are the diagonals of rhombi (equal-side parallelograms), hence at right angles and bisecting the center of mass angles of scattering.

Chapter 14

Electromagnetic Radiation

Electromagnetic radiation is a transverse energy wave that is composed of an oscillating electric field component, *E*, and an oscillating magnetic field component, *M*. The electric and magnetic fields are orthogonal to each other, and they are orthogonal to the direction of propogation of the wave. A wave is described by the wavelength, λ, which is the physical length of one complete oscillation, and the frequency, ν, which is the number of oscillations per second. The figure shows one wavelength of a wave of light. The names we give electromagnetic radiation for different wavelength and frequency ranges are listed in the electromagnetic spectrum document.

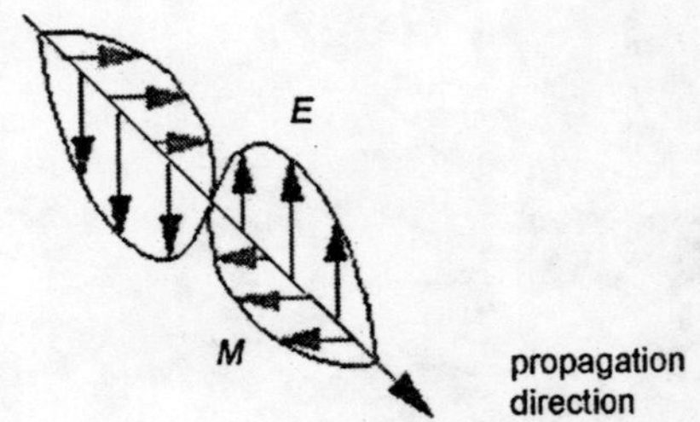

Fig. Schematic of an electromagnetic wave

VELOCITY OF LIGHT

Electromagnetic waves travel through a vacuum at a constant velocity of 2.99792×10^8 m/s, which is known as the speed of light, c. The relationship between the speed of light, wavelength, and frequency is:

$$c = \lambda\nu$$

When light passes through other media, the velocity of light decreases. For a given frequency of light, the wavelength also must decrease. This decrease in velocity is quantitated by the refractive index, n, which is the ratio of c to the velocity of light in another medium, v:

$$n = c / v$$

Since the velocity of light is lower in other media than in a vacuum, n is always a number greater than one. The table lists the refractive index of several examples. Refractive index is an intrinsic physical property of a substance, and can be used to monitor purity or the concentration of a solute in a solution. The refractive index of a material is measured with a refractometer, and is usually made versus air. If the precision warrants, the measurements can be corrected for vacuum. Note that the difference between n_{air} and n_{vacuum} is only significant in the fourth decimal place.

For anisotropic materials, such as quartz crystals, light of different polarizations (see below) will experience different refractive indices. These indices are called the ordinary refractive index, n_o, and the extraordinary refractive index, n_e.

medium	n*
air	1.0003
water	1.333
50per cent sucrose in water	1.420
carbon disulfide	1.628
crystalline quartz	1.544 (n_o)
1.553 (n_e)	
diamond	2.417

*measured with 589.3 nm light

POLARIZATION

An incoherent light source, such as the hot filament of a light bulb, consists of multiple, randomly oriented light emitters, which produce electromagnetic waves with their electric-field vector oriented in all directions. The resulting light emission is called unpolarized light.

Linearly (or plane) polarized light is light in which the electric-field vector is oscillating in only one direction. Linearly polarized light is produced by isolating one orientation of the electric field with a polarizer, or from lasers that contain polarized optical components.

Circularly polarized light is light in which the electric field vector is rotating around the axis of light propogation. The electric field vector can rotate in either the right or left direction (as viewed in the direction of light propogation), and the light is called right circularly polarized or left circularly polarized, respectively.

WAVE-PARTICLE DUALITY

Electromagnetic radiation shows both wave and particle characteristics depending on how the radiation is observed. Einstein first postulated that the energy of radiation is quantized and that radiation is composed of energy packets that were later named photons. The energy, E, of one photon depends on its frequency (or wavelength):

$$E = h\nu = hc/\lambda$$

where h is Planck's constant (6.62618×10^{-34} Js), ν is the frequency of the radiation, c is the speed of light, and λ is wavelength.

DE BROGLIE EQUATION

Moving particles; such as electrons, protons, and neutrons; have wave properties as described by the de Broglie equation:

$$\lambda = h / p$$

where λ is wavelength, h is Planck's constant, and p is the momentum of the particle. Beams of particles can therefore show wave effects such as interference.

ELECTROMAGNETIC SPECTRUM

VISIBLE SPECTRUM

The Visible Spectrum

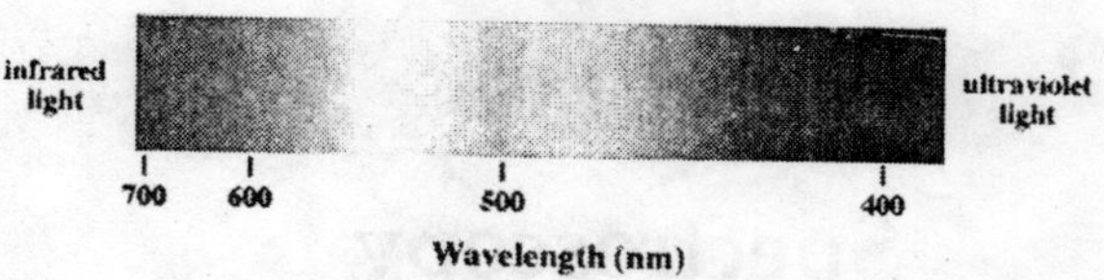

Figure. Electromagnetic Spectrum

Type of Radiation	Frequency Range (Hz)	Wavelength Range	Type of Transition
gamma-rays	10^{20}-10^{24}	<10^{-12} m	Nuclear
x-rays	10^{17}-10^{20}	1 nm-1 pm	inner electron
ultraviolet	10^{15}-10^{17}	400 nm-1 nm	outer electron
visible	4-7.5x10^{14}	750 nm-400 nm	outer electron
near-infrared	1x10^{14}-4x10^{14}	2.5 um-750 nm	outer electron molecular vibrations
infrared	10^{13}-10^{14}	25 um-2.5 um	molecular vibrations
microwaves	3x10^{11}-10^{13}	1 mm-25 um	molecular rotations, electron spin flips*
radio waves	<3x10^{11}	>1 mm	nuclear spin flips*

*energy levels split by a magnetic field

Chapter 15

Spectroscopy

Electron spectroscopies analyse the electrons that are ejected from a material for qualitative or semi-quantitative analysis. In general an excitation source such as x-rays or electrons will eject an electron from an inner-shell orbital of an atom. Detecting photoelectrons that are ejected by x-rays is call x-ray photoelectron spectroscopy (XPS) or electron spectroscopy for chemical analysis (ESCA). Detecting electrons that are ejected from higher orbitals to conserve energy during electron transitions is called Auger electron spectroscopy (AES). These electron processes are described below. Ejected electrons can escape only from a depth of approximately 3 nm or less, making electron spectroscopy most useful to study surfaces of solid materials. Depth profiling is accomplished by combining an electron spectroscopy with a sputtering source that removes surface layers.

AUGER ELECTRON SPECTROSCOPY (AES)

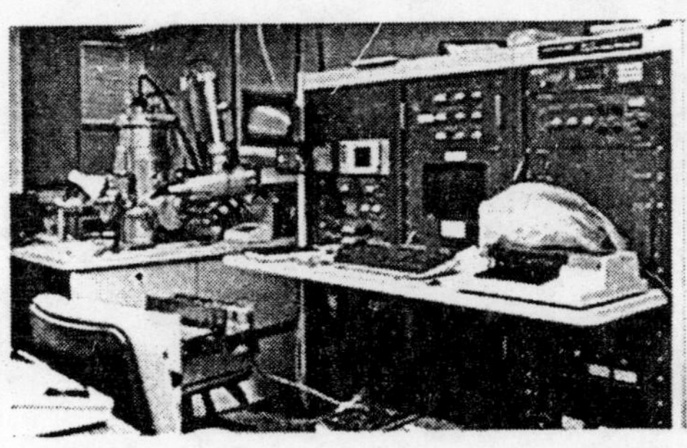

Figur. Auger electron spectrometer

Auger (pronounced ~o-jay) electron spectroscopy is an electron spectroscopic method that uses a beam of electrons to knock electrons out of inner-shell orbitals. Auger electrons are ejected to conserve energy when electrons in higher shells fill the vacancy in the inner shell. These Auger electrons have energies characteristic of the emitting atom due to the characteristic energy-level structure of that element.

X-RAY PHOTOELECTRON SPECTROSCOPY (XPS, ESCA)

X-ray photoelectron spectroscopy (XPS, also called electron spectroscopy for chemical analysis, ESCA) is a electron spectroscopic method that uses x-rays to eject electrons from inner-shell orbitals. The kinetic energy, E_k, of these photoelectrons is determined by the energy of the x-ray radiation, h,ν and the electron binding energy, E_b, as given by:

$$E_k = h\nu - E_b$$

The experimentally measured energies of the photoelectrons are given by:

$$E_k = h\nu - E_b - E_w$$

where E_w is the work function of the spectrometer.

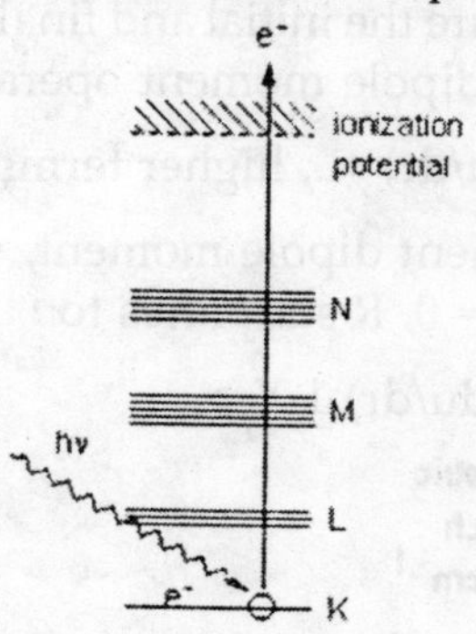

Fig. Energy-level diagram for XPS

The electron binding energies are dependent on the chemical environment of the atom, making XPS useful to identify the oxidation state and ligands of an atom.

XPS instruments consist of an x-ray source, an energy analyser for the photoelectrons, and an electron detector. The analysis and detection of photoelectrons requires that the sample be placed in a high-vacuum chamber. Since the photoelectron energy depends on x-ray energy, the excitation source must be monochromatic. The energy of the photoelectrons is analysed by an electrostatic analyser, and the photoelectrons are detected by an electron multiplier tube or a multichannel detector such as a microchannel plate.

INFRARED ABSORPTION SPECTROSCOPY (IR)

IR spectroscopy is the measurement of the wavelength and intensity of the absorption of mid-infrared light by a sample. Mid-infrared light (2.5 - 50 μm, 4000 - 200 cm^{-1}) is energetic enough to excite molecular vibrations to higher energy levels. The wavelength of IR absorption bands are characteristic of specific types of chemical bonds, and IR spectroscopy finds its greatest utility for identification of organic and organometallic molecules.

MECHANISM OF IR ABSORPTION

The transition moment for infrared absorption is:

$$R = < X_i \mid u \mid X_j \, dt >$$

where X_i and X_j are the initial and final states, respectively, and u is the electric dipole moment operator:

$$u = u_o + (r-r_e)(du/dr) + \ldots \text{ higher terms.}$$

u_o is the permanent dipole moment, which is a constant, and since $< X_i \mid X_j > = 0$, R simplifies to:

$$R = < X_i \mid (r-r_e)(du/dr) \mid X_j >$$

symmetric stretch $1340\ cm^{-1}$ — asymmetric stretch $2350\ cm^{-1}$

O=C=O
O = C = O ← equilibrium position → O = C = O
O = C = O

O = C=O
O=C = O

Fig. Examples of infrared active and inactive absorption bands in CO_2

The result is that there must be a change in dipole moment during the vibration for a molecule to absorb infrared radiation.

There is no change in dipole moment during the symmetric stretch vibration and the 1340 cm^{-1} band is not observed in the infrared absorption spectrum (the symmetric stretch is called infrared inactive). There is a change in dipole moment during the asymmetric stretch and the 2350 cm^{-1} band does absorb infrared radiation (the asymmetric stretch in infrared active). A related vibrational spectroscopic method is Raman spectroscopy, which has a different mechanism and therefore provides complementary information to infrared absorption.

INFRARED ABSORPTION BANDS

IR absorption spectroscopy uses mid-infrared light (2.5 - 50 μm, 4000 - 200 cm^{-1}) to detect specific types of chemical bonds in a sample for identification of organic and organometallic molecules.

Table of characteristic IR bands

Group	Bond	Appox. Energy (cm^{-1})
hydroxyl	O-H	3610-3640
amines	N-H	3300-3500
aromatic rings	C-H	3000-3100
alkenes	C-H	3020-3080
alkanes	C-H	2850-2960
nitriles	C=-N	2210-2260
carbonyl	C=O	1650-1750
amines	C-N	1180-1360

INFRARED ABSORPTION SPECTROMETERS

This document describes dispersive and Fourier-transform spectrometers that are used in infrared absorption spectroscopy.

DISPERSIVE INFRARED SPECTROMETERS

Common light sources are tungsten lamps, Nernst glowers, or glowbars. Dispersive IR spectrometers use a grating monochromator to select wavelengths and are commonly used when a single wavelength is desired to monitor the kinetics of a reaction or as a GC or LC detector.

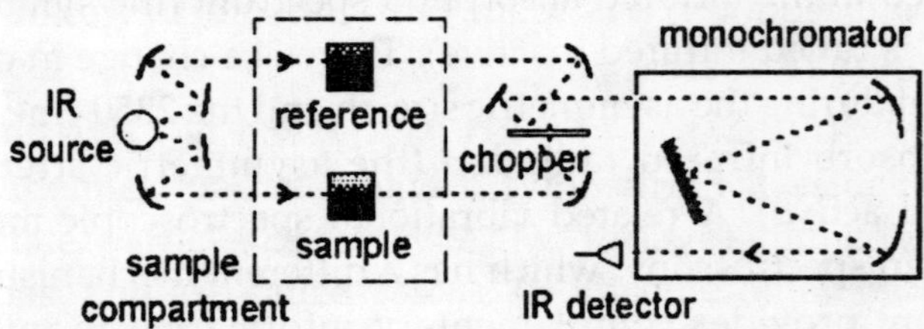

Fig. Figure of a dispersive IR absorption spectrometer

FOURIER-TRANSFORM INFRARED (FTIR)

Modern IR instruments more commonly use Fourier-transform techniques with a Michelson interferometer.

Fig. Figure an FTIR absorption spectrometer

CAVITY-RINGDOWN LASER ABSORPTION SPECTROSCOPY (CRLAS)

Cavity-ringdown laser absorption spectroscopy (CRLAS) is an ultrasensitive method to make quantitative absorption measurements of very low concentrations of analytes. The technique uses a laser pulse that is reflected back and forth between two highly reflecting mirrors. This procedure results in a very long path length, to which the measured absorbance is directly proportional as described by the Beer-Lambert law.

An optical detector is placed behind one of the mirrors to detect the small amount of the light that passes through the mirror. With no absorbing analyte present, the laser pulse will decrease in intensity after each round trip due to the loss of light through the monitoring mirror and other losses. When an absorbing species is present between the mirrors, the intensity of the laser pulse decreases more rapidly. The analyte concentration is determined by calibrating this decay time with known concentration of analyte.

CRLAS can be used from the near-UV to the mid-IR. The wavelength range is only limited by instrumental constraints, such as the availability of high reflectivity mirrors and suitable pulsed laser sources. Accessing the near and mid-IR is accomplished with frequency-conversion techniques such as Raman shifting or optical parametric oscillators.

In principle, the analyte can be in a solid, liquid, or gas. In practice, CRALS is primarily used to measure gas-phase species due to scattering losses in solids or liquids.

INTRACAVITY-ABSORPTION SPECTROSCOPY

Intracavity-absorption spectroscopy is an ultrasensitive method for measuring very small absorptions. Some examples are very low concentrations of analytes, and very weak absorption such as infrared overtone bands. The technique takes advantage of the very high light intensities within a laser resonance cavity, and the sensitivity of the lasing output to losses within the cavity. Small losses within the laser cavity due to absorption results in large changes in the intensity of the laser output.

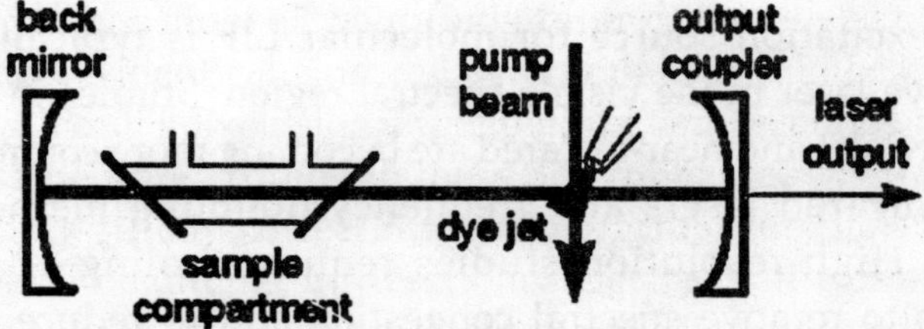

Fig. Schematic of an intracavity-absorption experiment

To record a spectrum the laser output must be tunable, therefore the most common lasers used in intracavity-absorption measurements are dye lasers.

Samples can be solids, liquids, or gases. Liquids and gases are contained in a sample holder that has windows at Brewster's angle to minimize losses due to reflection.

Small absorptions are detected by monitoring the output of the laser beam with a detector. Since the laser light is relatively intense, the main requirement for the detector is that it not be destroyed by the laser beam.

LASER-INDUCED FLUORESCENCE (LIF)

Laser-induced fluorescence (LIF) is the optical emission from molecules that have been excited to higher energy levels by absorption of electromagnetic radiation. The main advantage of fluorescence detection compared to absorption measurements is the greater sensitivity achievable because the fluorescence signal has a very low background. For molecules that can be resonant excitated, LIF provides selective excitation of the analyte to avoid interferences. LIF is useful to study the electronic structure of molecules and to make quantitative measurements of analyte concentrations. Analytical applications include monitoring gas-phase concentrations in the atmosphere, flames, and plasmas; and remote sensing using light detection and ranging (LIDAR). Because of the differences in the nature of the energy-level structure between atoms and molecules, the discussion on atomic fluorescence spectroscopy is in a separate document.

The excitation source for molecular LIF is typically a tunable dye laser in the visible spectral region. Studies in the near-ultraviolet and near-infrared are becoming more common as near-infrared lasers and frequency-doubling methods improve. High-resolution studies require cooling of the molecules to remove spectral congestion and to reduce the Doppler width of the transitions. A separate document on high-resolution spectroscopy describes cooling methods such as molecular beams, free-jet expansions, and cryogenic glass or crystalline matrices.

RAMAN SPECTROSCOPY

Raman spectroscopy is the measurement of the wavelength and intensity of inelastically scattered light from molecules. The Raman scattered light occurs at wavelengths that are shifted from the incident light by the energies of molecular vibrations. The mechanism of Raman scattering is different from that of infrared absorption, and Raman and IR spectra provide complementary information. Typical applications are in structure determination, multicomponent qualitative analysis, and quantitative analysis.

THEORY

The Raman scattering transition moment is:

$R = < X_i \mid a \mid X_j >$

where X_i and X_j are the initial and final states, respectively, and a is the polarizability of the molecule:

$a = a_o + (r-r_e)(da/dr) +...$ higher terms

where r is the distance between atoms and a_o is the polarizability at the equilibrium bond length, r_e. Polarizability can be defined as the ease of which an electron cloud can be distorted by an external electric field. Since a_o is a constant and $< X_i \mid X_j > = 0$, R simplifies to:

$R = < X_i \mid (r-r_e)(da/dr) \mid X_j >$

The result is that there must be a change in polarizability during the vibration for that vibration to inelastically scatter radiation.

Examples of Raman active and inactive vibrations in CO_2

The polarizability depends on how tightly the electrons are bound to the nuclei. In the symmetric stretch the strength of electron binding is different between the minimum and maximum internuclear distances. Therefore the polarizability changes during the vibration and this vibrational mode scatters Raman light (the vibration is Raman active). In the asymmetric stretch the electrons are more easily polarized in the bond that expands but are less easily polarized in the bond that compresses. There is no overall change in polarizability and the asymmetric stretch is Raman inactive.

Raman line intensities are proportional to:

nu^4 * sigma(nu) * I * exp($-E_i$/kT) * C

where nu is the frequency of the incident radiation, sigma(nu) is the Raman cross section (typically 10^{-29} cm^2), I is the radiation intensity, exp($-E_i$/kT) is the Boltzmann factor for state i, and C is the analyte concentration.

The most common light source in Raman spectroscopy is an Ar-ion laser. Resonance Raman spectroscopy requires tunable radiation and sources are Ar-ion-laser-pumped cw dye lasers, or high-repetition-rate excimer-laser-pumped pulsed dye lasers. Because Raman scattering is a weak process, a key requirement to obtain Raman spectra is that the spectrometer provide a high rejection of scattered laser light. New methods such as very narrow rejection filters and Fourier-transform techniques are becoming more widespread.

RESONANCE-IONIZATION SPECTROSCOPY (RIS)

Resonance-ionization spectroscopy (RIS) is a spectroscopic method that uses resonant laser excitation to promote an atom or molecule above its ionization potential to create an ion. The ions are detected as a current by a biased collector or by a mass spectrometer. RIS is useful to study the electronic structure of atoms or molecules and to make quantitative measurements of analyte concentrations. For more information see the document on RIMS.

LASER-ENHANCED IONIZATION

Laser-enhanced ionization (LEI) ionizes atoms in a flame. It relies on resonant laser excitation of an atom to a higher electronic excited state. The excited atom is then ionized by collisions in the flame. The ions are collected by biased wires or plates in or near the flame. The LEI method produces lower limits of detection for several elements.

NEAR-INFRARED ABSORPTION SPECTROSCOPY (NIR)

NIR spectroscopy is the measurement of the wavelength and intensity of the absorption of near-infrared light by a sample. Near-infrared light spans the 800 nm - 2.5 μm (12,500 - 4000 cm^{-1}) range and is energetic enough to excite overtones and combinations of molecular vibrations to higher energy levels. NIR spectroscopy is typically used for quantitative measurement of organic functional groups, especially O-H, N-H, and C=O. Detection limits are typically 0.1per cent and applications include pharmaceutical, agricultural, polymer, and clinical analysis.

The components and design of NIR instrumentation are similar to uv-vis absorption spectrometers. The light source is usually a tungsten lamp and the detector is usually a PbS solid-state detector. Sample holders can be glass or quartz and typical solvents are CCl_4 and CS_2. The convenient instrumentation of NIR spectroscopy compared to IR spectroscopy makes it much more suitable for on-line monitoring and process control.

POLARIMETRY

Anisotropic crystalline solids, and samples containing an excess of one enantiomer of a chiral molecule, can rotate the orientation of plane-polarized light. Such substances are said to have optical activity. Measurement of this change in polarization orientation is called polarimetry, and the measuring instrument is called a polarimeter. These measurements are useful for studying the structure of anisotropic materials, and for checking the purity of chiral mixtures. A sample that contains only one enantiomer of a chiral molecule is said to be optically pure. The enantiomer that rotates light to the right, or clockwise when viewing in the direction of light propagation, is called the dextrorotatory (d) or (+) enantiomer, and the enantiomer that rotates light to the left, or counterclockwise, is called the levorotatory (l) or (-) enantiomer.

Optical rotation occurs because optically active samples have different refractive indices for left- and right-circularly polarized light. Another way to make this statement is that left- and right-circularly polarized light travel through an optically active sample at different velocities. This condition occurs because a chiral center has a specific geometric arrangement of four different substituents, each of which has a different electronic polarizability. Light travels through matter by interacting with the electron clouds that are present. Left-circularly polarized light therefore interacts with an anisotropic medium differently than does right-circularly polarized light.

Linearly or plane-polarized light is the superposition of equal intensities of left- and right-circularly polarized light. As plane-polarized light travels through an optically active sample, the left- and right-circularly polarized components travel at different velocities. This difference in velocities creates a phase shift between the two circularly polarized components when they exit the sample. Summing the two components still produces linearly polarized light, but at a different orientation from the light entering the sample.

The simplest polarimeter consists of a monochromatic light source, a polarizer, a sample cell, a second polarizer, which is called the analyser, and a light detector. The analyser is oriented 90° to the polarizer so that no light reaches the detector.

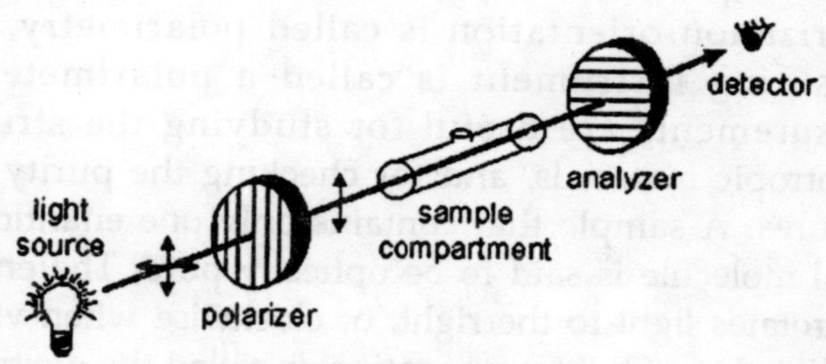

Fig. Schematic of a polarimeter

When an optically active substance is present in the beam, it rotates the polarization of the light reaching the analyser so that there is a component that reaches the detector. The angle that the analyser must be rotated to return to the minimum detector signal is the optical rotation, α.

The amount of optical rotation depends on the number of optically active species through which the light passes, and thus depends on both the sample path length and the analyte concentration. Specific rotation, [α], provides a normalize quantity to correct for this dependence, and is defined as:

$$[\alpha] = \frac{\alpha}{1 + d}$$

where α is the measured optical rotation in degrees, l is the sample path length in decimeters (dm), and d is the density if the sample is pure liquid, or the concentration if the sample is a solution. In either case, the units of d are g/cm^3.

The optical rotatory dispersion (ORD) is the optical rotation as a function of wavelength. It is recorded using a spectropolarimeter, which has a tungsten lamp and a scanning monochromator as the light source. A motorized mount rotates the analyser to maintain a minimum signal at the detector. Usually a modulation is introduced into the polarization angle of the light beam, so that a DC signal to the analyser motor then keeps the detector signal centered at the minimum value.

OPTICAL MATERIALS

This document provides data on materials that are used for optical components (mirrors, lenses, and windows) in different parts of the electromagnetic spectrum.

MIRRORS

X-rays	
Ultraviolet	aluminium
Visible	aluminium
Near infrared	gold
infrared	copper, gold

LENSES

X-rays	
Ultraviolet	fused silica (synthetic quartz), sapphire

Visible	glass, sapphire
Near infrared	glass, sapphire
infrared	CaF_2, ZnSe

WINDOWS

X-rays	beryllium
Ultraviolet	fused silica, sapphire
Visible	glass, sapphire
Near infrared	glass, sapphire
infrared	NaCl, BaF_2, CaF_2, ZnSe

Refractive index of some optical materials

medium	n*
air	1.0003
water	1.333
CaF_2	1.434
BaF_2	1.474
glass	1.5-1.9
synthetic quartz (fused silica)	1.458
crystalline quartz	1.544 (n_o), 1.553 (n_e)
calcite	1.658 (n_o), 1.486 (n_e)
sapphire	1.769
diamond	2.417
silicon	3.478 (1.55 μm)
ZnSe	2.624
	2.403 (10.6 μm)

*Refractive index at 589.3 nm and room temperature, unless noted otherwise.

Chapter 16

Mass Spectrometry

INTRODUCTION

Mass spectrometers use the difference in mass-to-charge ratio (m/e) of ionized atoms or molecules to separate them from each other. Mass spectrometry is therefore useful for quantitation of atoms or molecules and also for determining chemical and structural information about molecules. Molecules have distinctive fragmentation patterns that provide structural information to identify structural components.

The general operation of a mass spectrometer is:

1. Create gas-phase ions
2. Separate the ions in space or time based on their mass-to-charge ratio
3. Measure the quantity of ions of each mass-to-charge ratio

The ion separation power of a mass spectrometer is described by the resolution, which is defined as:

$R = m / \Delta m$,

where Δm is the ion mass and m is the difference in mass between two resolvable peaks in a mass spectrum. E.g., a mass spectrometer with a resolution of 1000 can resolve an ion with a m/e of 100.0 from an ion with an m/e of 100.1.

In general a mass spectrometer consists of an ion source, a mass-selective analyser, and an ion detector. Since mass spectrometers create and manipulate gas-phase ions, they operate in a high-vacuum system. The magnetic-sector, quadrupole, and time-of-flight designs also require extraction and acceleration ion optics to transfer ions from the source region into the mass analyser. The details of mass analyser designs are discussed in the individual documents listed below. Basic descriptions of sample introduction/ionization and ion detection are discussed in separate documents on ionization methods and ion detectors, respectively.

MASS ANALYSER DESIGNS:

- Fourier-transform MS
- Ion-trap MS
- Magnetic-sector MS
- Quadrupole MS
- Time-of-flight MS

MASS SPECTROMETRY IONIZATION METHODS

CHEMICAL IONIZATION (CI)

CI uses a reagent ion to react with the analyte molecules to form ions by either a proton or hydride transfer:

$$MH + C_2H_5^+ \rightarrow MH_2^+ + C_2H_4$$

$$MH + C_2H_5^+ \rightarrow M^+ + C_2H_6$$

The reagent ions are produced by introducing a large excess of methane (relative to the analyte) into an electron impact (EI) ion source. Electron collisions produce CH_4^+ and CH_3^+ which further react with methane to form CH_5^+ and $C_2H_5^+$:

$$CH_4^+ + CH_4 \rightarrow CH_5^+ + CH_3$$

$$CH_3^+ + CH_4 \rightarrow C_2H_5^+ + H_2$$

PLASMA AND GLOW DISCHARGE

A plasma is a hot, partially-ionized gas that effectively excites and ionizes atoms.

A glow discharge is a low-pressure plasma maintained between two electrodes. It is particularly effective at sputtering and ionizing material from solid surfaces.

ELECTRON IMPACT (EI)

An EI source uses an electron beam, usually generated fron a tungsten filament, to ionize gas-phase atoms or molecules. An electron from the beam knocks an electron off of analyte atoms or molecules to create ions.

ELECTROSPRAY IONIZATION (ESI)

The ESI source consists of a very fine needle and a series of skimmers. A sample solution is sprayed into the source chamber to form droplets. The droplets carry charge when the exit the capillary and, as the solvent evapourates, the droplets disappear leaving highly charged analyte molecules. ESI is particularly useful for large biological molecules that are difficult to vapourize or ionize.

FAST-ATOM BOMBARDMENT (FAB)

In FAB a high-energy beam of netural atoms, typically Xe or Ar, strikes a solid sample causing desorption and ionization. It is used for large biological molecules that are difficult to get into the gas phase. FAB causes little fragmentation and usually gives a large molecular ion peak, making it useful for molecular weight determination.

The atomic beam is produced by accelerating ions from an ion source though a charge-exchange cell. The ions pick up an electron in collisions with netural atoms to form a beam of high energy atoms.

FIELD IONIZATION

Molecules can lose an electron when placed in a very high electric field. High fields can be created in an ion source by applying a high voltage between a cathode and an anode called a field emitter. A field emitter consists of a wire covered with microscopic carbon dendrites, which greatly amplify the effective field at the carbon points.

LASER IONIZATION (LIMS)

A laser pulse ablates material from the surface of a sample, and creates a microplasma that ionizes some of the sample constituents. The laser pulse accomplishes both vapourization and ionization of the sample.

MATRIX-ASSISTED LASER DESORPTION IONIZATION (MALDI)

MALDI is a LIMS method of vapourizing and ionizing large biological molecules such as proteins or DNA fragments. The biological molecules are dispersed in a solid matrix such as nicotinic acid.

A UV laser pulse ablates the matrix which carries some of the large molecules into the gas phase in an ionized form so they can be extracted into a mass spectrometer.

PLASMA-DESORPTION IONIZATION (PD)

Decay of ^{252}Cf produces two fission fragments that travel in opposite directions. One fragment strikes the sample knocking out 1-10 analyte ions. The other fragment strikes a detector and triggers the start of data acquisition. This ionization method is especially useful for large biological molecules.

RESONANCE IONIZATION (RIMS)

One or more laser beams are tuned in resonance to transitions of a gas-phase atom or molecule to promote it in a stepwise fashion above its ionization potential to create an ion. Solid samples must be vapourized by heating, sputtering, or laser ablation.

SECONDARY IONIZATION (SIMS)

A primary ion beam; such as $^{3}He^{+}$,$^{16}O^{+}$, or $^{40}Ar^{+}$; is accelerated and focused onto the surface of a sample and sputters material into the gas phase. Approximately 1per cent of the sputtered material comes off as ions, which can then be analysed by a mass spectrometer. SIMS has the advantage that material can be continually sputtered from a surface to determine analyte concentrations as a function of distance from the original surface (depth profiling).

SPARK SOURCE

A spark source ionizes analytes in solid samples by pulsing an electric current across two electrodes. If the sample is a metal it can serve as one of the electrodes, otherwise it can be mixed with graphite and placed in a cup-shaped electrode.

THERMAL IONIZATION (TIMS)

Thermal ionization is used for elemental or refractory materials. A sample is deposited on a metal ribbon, such as Pt or Re, and an electric current heats the metal to a high temperature. The ribbon is often coated with graphite to provide a reducing effect.

ION DETECTORS

CHANNELTRON

A channeltron is a horn-shaped continuous dynode structure that is coated on the inside with a electron emissive material. An ion striking the channeltron creates secondary electrons that have an avalanche effect to create more secondary electrons and finally a current pulse.

DALY DETECTOR

A Daly detector consists of a metal knob that emits secondary electrons when struck by an ion. The secondary electrons are accelerated onto a scintillator that produces light that is then detected by a photomultiplier tube.

ELECTRON MULTIPLIER TUBE (EMT)

Electron multiplier tubes are similar in design to photomultiplier tubes. They consist of a series of biased dynodes that eject secondary electrons when they are struck by an ion. They therefore multiply the ion current and can be used in analog or digital mode.

FARADAY CUP

A Faraday cup is a metal cup that is placed in the path of the ion beam. It is attached to an electrometer, which measures the ion-beam current. Since a Faraday cup can only be used in an analog mode it is less sensitive than other detectors that are capable of operating in pulse-counting mode.

MICROCHANNEL PLATE

A microchannel plate consists of an array of glass capillaries (10-25 um inner diameter) that are coated on the inside with a electron-emissive material. The capillaries are biased at a high voltage and like the channeltron, an ion that strikes the inside wall one of the capillaries creates an avalanche of secondary electrons. This cascading effect creates a gain of 10^3 to 10^4 and produces a current pulse at the output.

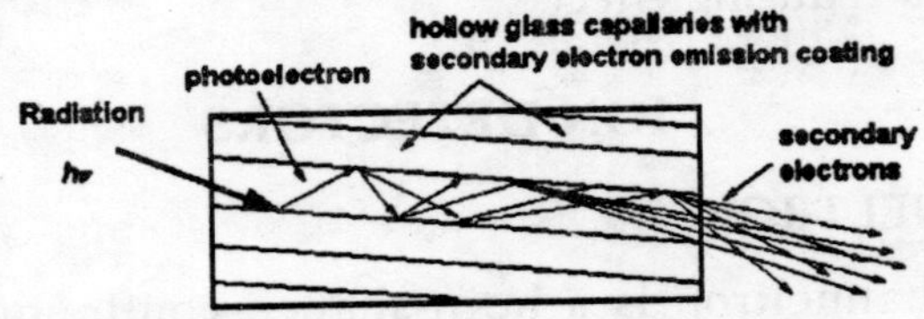

Figure. Schematic of a microchannel plate

Microchannel plates (MCP) are also used as an intensifier for low-intensity light detection with array detectors.

FOURIER-TRANSFORM MASS SPECTROMETRY

Fourier-transform mass spectrometry takes advantage of ion-cyclotron resonance to select and detect ions.

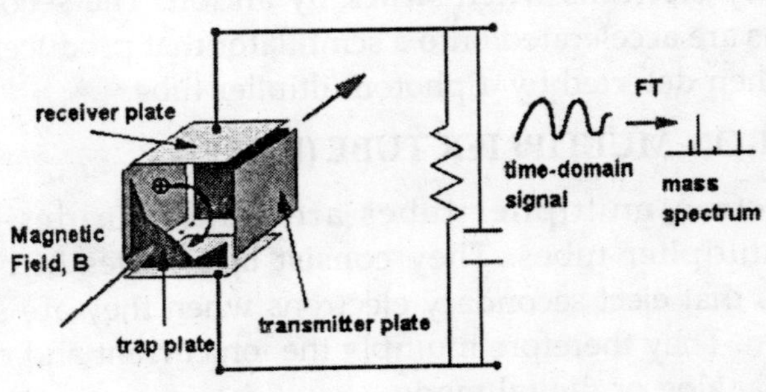

Fig. Schematic of a FT-MS

ION-TRAP MASS SPECTROMETRY

The ion-trap mass spectrometer uses three electrodes to trap ions in a small volume. The mass analyser consists of a ring electrode separating two hemispherical electrodes. A mass spectrum is obtained by changing the electrode voltages to eject the ions from the trap. The advantages of the ion-trap mass spectrometer include compact size, and the ability to trap and accumulate ions to increase the signal-to-noise ratio of a measurement.

MAGNETIC-SECTOR MASS SPECTROMETRY

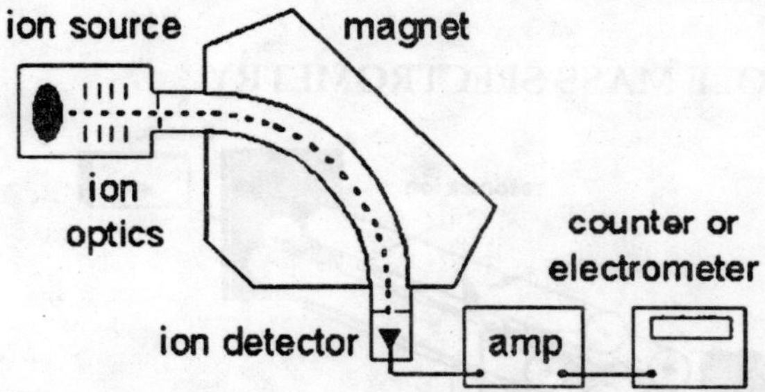

Fig. Schematic of a magnetic-sector mass spectrometer

THEORY

The ion optics in the ion-source chamber of a mass spectrometer extract and accelerate ions to a kinetic energy given by:

$$K.E. = 0.5\ mv^2 = eV$$

where m is the mass of the ion, v is it's velocity, e is the charge of the ion and V is the applied voltage of the ion optics.

The ions enter the flight tube between the poles of a magnet and are deflected by the magnetic field, H. Only ions of mass-to-charge ratio that have equal centrifugal and centripetal forces pass through the flight tube:

$$mv^2 / r = Hev$$

centrifugal = centripetal forces.

Where r is the radius of curvature of the ion path:

$$r = mv / eH$$

This equation shows that the m/e of the ions that reach the detector can be varied by changing either H or V.

Single Focusing analyzers:

A circular beam path of 180, 90, or 60 degrees can be used. The various forces influencing the particle separate ions with different mass-to-charge ratios.

Double Focusing analyzers:

An electrostatic analyzer is added in this type of instrument to separate particles with difference in kinetic energies.

QUADRUPOLE MASS SPECTROMETRY

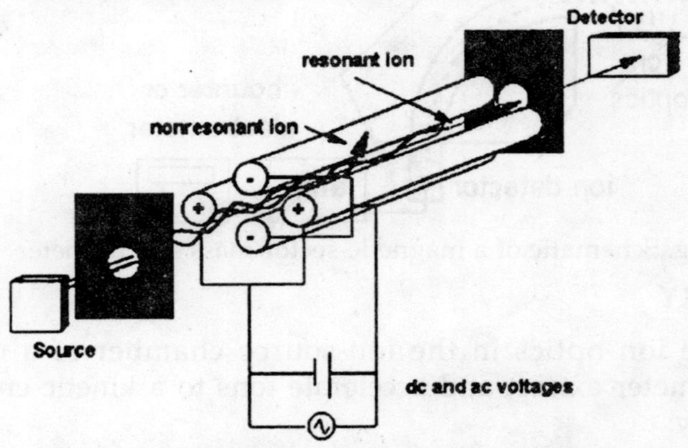

Fig. Schematic of a quadrupole filter

A quadrupole mass filter consists of four parallel metal rods arranged as in the figure below. Two opposite rods have an applied potential of (U+Vcos(wt)) and the other two rods have a potential of -(U+Vcos(wt)), where U is a dc voltage and Vcos(wt) is an ac voltage. The applied voltages affect the trajectory of ions traveling down the flight path centered between the four rods. For given dc and ac voltages, only ions of a certain mass-to-charge ratio pass through the quadrupole filter and all other ions are thrown out of their original path. A mass spectrum is obtained by monitoring the ions passing through the quadrupole filter as the voltages on the rods are varied. There are two methods: varying w and holding U and V constant, or varying U and V (U/V) fixed for a constant w.

Quadrupole mass spectrometers consist of an ion source, ion optics to accelerate and focus the ions through an aperture into the quadrupole filter, the quadrupole filter itself with control voltage supplies, an exit aperture, an ion detector and electronics, and a high-vacuum system.

TIME-OF-FLIGHT MASS SPECTROMETRY (TOF-MS)

A time-of-flight mass spectrometer uses the differences in transit time through a drift region to separate ions of different masses. It operates in a pulsed mode so ions must be produced or extracted in pulses. An electric field accelerates all ions into a field-free drift region with a kinetic energy of qV, where q is the ion charge and V is the applied voltage. Since the ion kinetic energy is $0.5mv^2$, lighter ions have a higher velocity than heavier ions and reach the detector at the end of the drift region sooner.

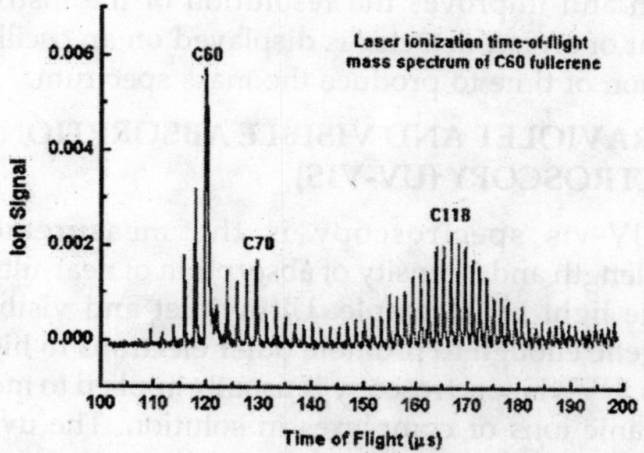

Fig. Example of a TOF mass spectrum

The mass of C60 is 720 amu (fullerene mass spectrum courtesy of Craig Watson, Virginia Tech).

THEORY

K.E. = qV

$^1/_2\, mv^2 = qV$

$V = (^{2qV}/_{m})^{1/2}$

The transit time (t) through the drift tube is L/V where L is the length of the drift tube.

$t = L / (^{2V}/_{m/q})^{1/2}$

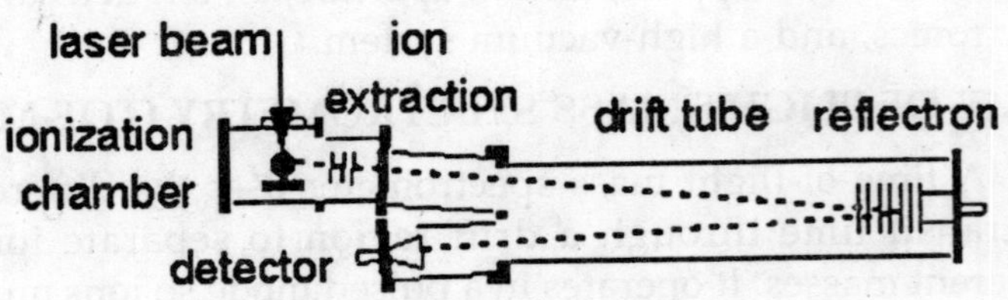

Fig. Schematic of a reflectron TOF-MS

This schematic shows ablation of ions from a solid sample with a pulsed laser. The reflectron is a series of rings or grids that act as an ion mirror. This mirror compensates for the spread in kinetic energies of the ions as they enter the drift region and improves the resolution of the instrument. The output of an ion detector is displayed on an oscilloscope as a function of time to produce the mass spectrum.

ULTRAVIOLET AND VISIBLE ABSORPTION SPECTROSCOPY (UV-VIS)

UV-vis spectroscopy is the measurement of the wavelength and intensity of absorption of near-ultraviolet and visible light by a sample. Ultraviolet and visible light are energetic enough to promote outer electrons to higher energy levels. UV-vis spectroscopy is usually applied to molecules and inorganic ions or complexes in solution. The uv-vis spectra have broad features that are of limited use for sample identification but are very useful for quantitative measurements. The concentration of an analyte in solution can be determined by measuring the absorbance at some wavelength and applying the Beer-Lambert Law.

The light source is usually a hydrogen or deuterium lamp for uv measurements and a tungsten lamp for visible measurements. The wavelengths of these continuous light sources are selected with a wavelength separator such as a prism or grating monochromator. Spectra are obtained by scanning the wavelength separator and quantitative measurements can be made from a spectrum or at a single wavelength.

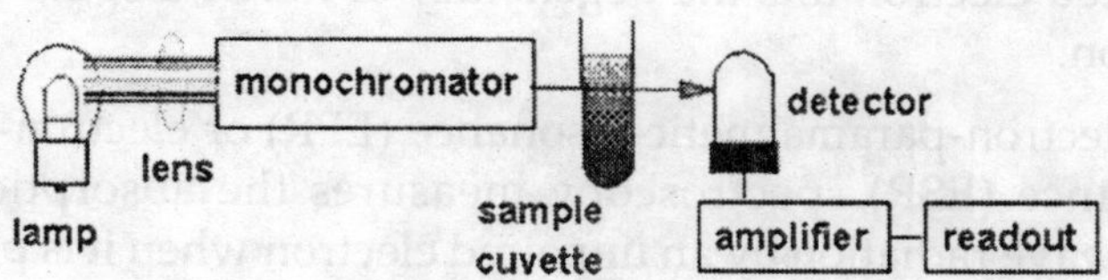

Fig. Schematic of a single beam uv-vis spectrophotometer

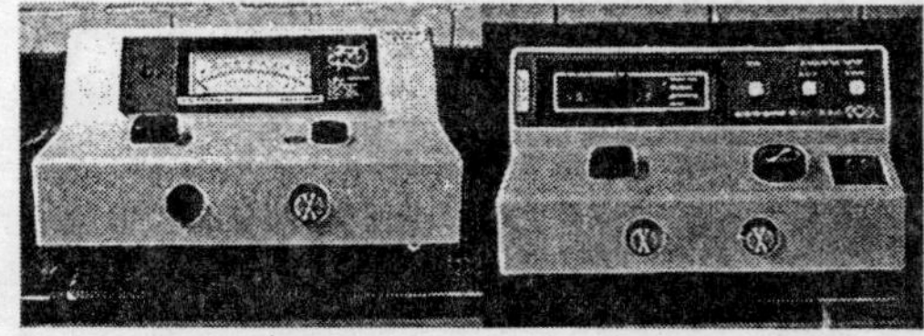

Fig. Pictures of single beam uv-vis spectrophotometers (Spectronic 20 and 20D)

Larger picture of the Spectronic 20D with labels.

DUAL-BEAM UV-VIS SPECTROPHOTOMETER

In single-beam uv-vis absorption spectroscopy, obtaining a spectrum requires manually measuring the transmittance of the sample and solvent at each wavelength. The double-beam design greatly simplifies this process by measuring the transmittance of the sample and solvent simultaneously. The detection electronics can then manipulate the measurements to give the absorbance.

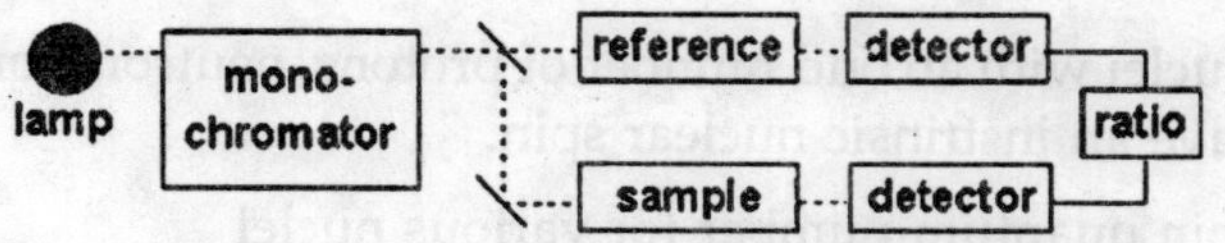

Fig. Schematic of a dual-beam uv-vis spectrophotometer

ELECTRON PARAMAGNETIC RESONANCE (EPR, ESR) SPECTROSCOPY

When an atom or molecule with an unpaired electron is placed in a magnetic field, the spin of the unpaired electron can align either in the same direction or in the opposite direction as the field. These two electron alignments have different energies and application of a magnetic field to an unpaired electron lifts the degeneracy of the ±1/2 spins of the electron.

Electron-paramagnetic-resonance (EPR) or electron-spin-resonance (ESR) spectroscopy measures the absorption of microwave radiation by an unpaired electron when it is placed in a strong magnetic field.

Species that contain unpaired electrons:

1. Free radicals
2. Odd electron molecules
3. Transition-metal complexes
4. Lanthanide ions
5. Triplet-state molecules

A klystron tube generates monochromatic microwave radiation (~9500 MHz) in an EPR instrument. The microwave radiation travels down a waveguide to the sample which is held between magnets.

Spectra are obtained by measuring the absorption of the microwave radiation while scanning the magnetic-field strength. EPR spectra are usually displayed in derivative form to improve the signal-to-noise ratio.

NUCLEAR MAGNETIC RESONANCE (NMR) SPECTROSCOPY

Nuclei with an odd number of protons, neutrons, or both, will have an instrinsic nuclear spin.

Spin quantum number for various nuclei

Number of protons	Number of Neutrons	Spin Quantum Number	Examples
Even	Even	0	^{12}C, ^{16}O, ^{32}S
Odd	Even	1/2	^{1}H, ^{19}F, ^{31}P
"	"	3/2	^{11}B,^{35}Cl, ^{79}Br

Even	Odd	1/2	^{13}C
"	"	3/2	^{127}I
"	"	5/2	^{17}O
Odd	Odd	1	^{2}H, ^{14}N

When a nucleus with a non-zero spin is placed in a magnetic field, the nuclear spin can align in either the same direction or in the opposite direction as the field. These two nuclear spin alignments have different energies and application of a magnetic field lifts the degeneracy of the nuclear spins. A nucleus that has its spin aligned with the field will have a lower energy than when it has its spin aligned in the opposite direction to the field.

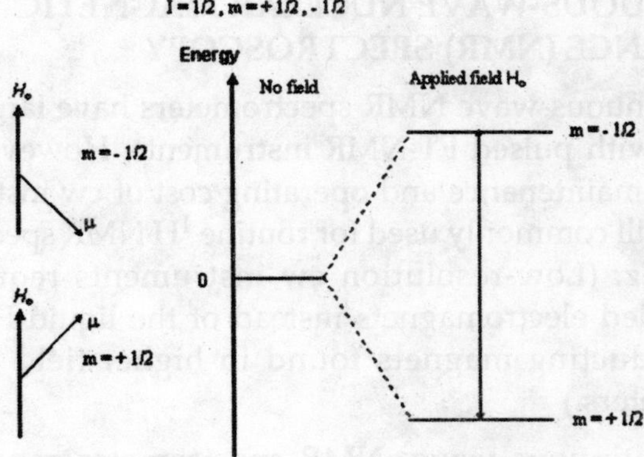

Nuclear magnetic resonance (NMR) spectroscopy is the absorption of radiofrequency radiation by a nucleus in a strong magnetic field. Absorption of the radiation causes the nuclear spin to realign or flip in the higher-energy direction. After absorbing energy the nuclei will reemit RF radiation and return to the lower-energy state.

The energy of a NMR transition depends on the magnetic-field strength and a proportionality factor for each nucleus called the magnetogyric ratio. The local environment around a given nucleus in a molecule will slightly perturb the local magnetic field exerted on that nucleus and affect its exact transition energy. This dependence of the transition energy on the position of a particular atom in a molecule makes NMR spectroscopy extremely useful for determining the structure of molecules.

There are two NMR spectrometer designs, continuous-wave (cw), and pulsed or Fourier-transform (FT-NMR). CW-NMR spectrometers have largely been replaced with pulsed FT-NMR instruments. However due to the lower maintenance and operating cost of cw instruments, they are still commonly used for routine ^{1}H NMR spectroscopy at 60 MHz. (Low-resolution cw instruments require only water-cooled electromagnets instead of the liquid-He-cooled superconducting magnets found in higher-field FT-NMR spectrometers.) These two spectrometer designs are described in separate CW-NMR and FT-NMR documents.

CONTINUOUS-WAVE NUCLEAR MAGNETIC RESONANCE (NMR) SPECTROSCOPY

Continuous-wave NMR spectrometers have largely been replaced with pulsed FT-NMR instruments. However due to the lower maintenance and operating cost of cw instruments, they are still commonly used for routine ^{1}H NMR spectroscopy at 60 MHz. (Low-resolution cw instruments require only water-cooled electromagnets instead of the liquid-He-cooled superconducting magnets found in higher-field FT-NMR spectrometers.)

A continuous wave -NMR spectrometer consists of a control console, magnet, and two orthogonal coils of wire that serve as antennas for radiofrequency (RF) radiation. One coil is attached to an RF generator and serves as a transmitter. The other coil is the RF pick-up coil and is attached to the detection electronics.

Since the two coils are orthogonal, the pick-up coil cannot directly receive any radiation from the generator coil. When a nucleus absorbs RF radiation, it can become reoriented due to its normal movement in solution and re-emit the RF radiation is a direction that can be received by the pick-up coil. This orthogonal coil arrangement greatly increases the sensitivity of NMR spectroscopy, similar to optical fluorescence.

Spectra are obtained by scanning the magnet and recording the pick-up coil signal on paper at the control console.

FOURIER-TRANSFORM NUCLEAR MAGNETIC RESONANCE (FT-NMR) SPECTROSCOPY

Fourier-transform NMR spectrometers use a pulse of radiofrequency (RF) radiation to cause nuclei in a magnetic field to flip into the higher-energy alignment. Due to the Heisenberg uncertainty principle, the frequency width of the RF pulse (typically 1-10 μs) is wide enough to simultaneously excite nuclei in all local environments. All of the nuclei will re-emit RF radiation at their respective resonance frequencies, creating an interference pattern in the resulting RF emission versus time, known as a free-induction decay (FID). The frequencies are extracted from the FID by a Fourier transform of the time-based data.

An FT-NMR spectrometer consists of a control console, magnet, and a coil of wire that serves as the antenna for transmitting and receiving the RF radiation. (Only one coil is necessary because signal reception does not begin until after the end of the excitation pulse.) Because the FID results from the emission due to nuclei in all environments, each pulse contains an interference pattern from which the complete spectrum can be obtained. Because of this multiplex (or Fellgett) advantage, repetitive signals can be summed and averaged to greatly improve the signal-to-noise ratio of the resulting FID.

MOSSBAUER SPECTROSCOPY

The Mossbauer effect is the recoil-free emission of gamma radiation from a solid radioactive material. Since the gamma emission is recoil-free, it can be resonantly absorbed by stationary atoms, i.e., also in a solid. The nuclear transitions are very sensitive to the local environment of the atom and Mossbauer spectroscopy is a sensitive probe of the different environments an atom occupies in a solid material.

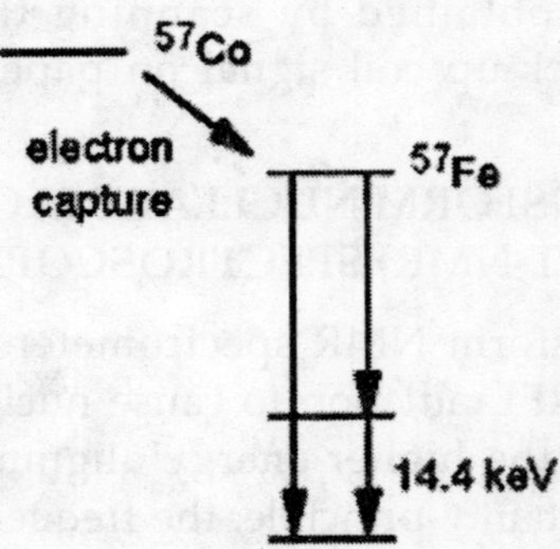

Fig. Production of gamma rays for ^{57}Fe Mossbauer spectroscopy

Approximately 90per cent of the ^{57}Fe nuclear excited state decays through the intermediate level to produce 14.4 keV gamma radiation. These gamma photons can then be absorbed by ^{57}Fe in a sample.

The gamma ray source is a radioactive element that is mechanically vibrated back and forth to Doppler shift the energy of the emitted gamma radiation. The schematic below shows a transmission Mossbauer experiment. As the energy of the gamma radiation is scanned by Doppler shifting, the detector records the frequencies of gamma radiation that are absorbed by the sample.

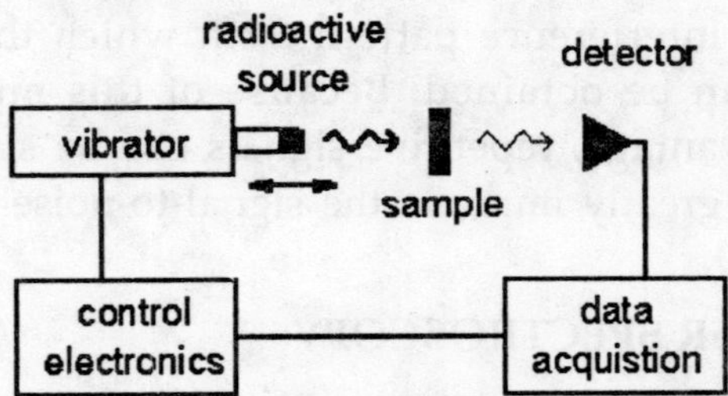

Fig. Schematic of an experimental set-up for transmission Mossbauer spectroscopy

X-RAY FLUORESCENCE

X-ray fluorescence is a spectroscopic method that is commonly used for solids in which secondary x-ray emission is generated by excitation of a sample with x-rays. The x-rays eject inner-shell electrons. Outer-shell electrons take their place and emit photons in the process. The wavelength of the photons depends on the energy difference between the outer-shell and inner-shell electron orbitals. The amount of x-ray fluorescence is very sample dependent and quatitative analysis requires calibration with standards that are similar to the sample matrix.

Solid samples are usually powdered and pressed into a wafer or fused in a borate glass. The sample is then placed in the sample chamber of an XRF spectrometer, and irradiated with a primary X-ray beam. The X-ray fluorescence is recorded with either an X-ray detector after wavelength dispersion or with an energy-dispersive detector.

RADIATION SOURCES FOR ABSORPTION AND FLUORESCENCE SPECTROSCOPY

LAMPS

Lamps convert electrical energy into radiation. Different designs and materials are needed to produce light in different parts of the electromagnetic spectrum. The following sections describe several different types of lamps that are useful in spectroscopy.

BLACKBODY SOURCES

A hot material, such as an electrically-heated filament in a light bulb, emits a continuum spectrum of light. The spectrum is approximated by Planck's radiation law for blackbody radiators:

$$B = \frac{8\pi h v^3}{c^3\left(e^{-hv/kT} - 1\right)}$$

The most common incandescent lamps and their wavelength ranges are:

tungsten filament lamps: 350 nm - 2.5 mm

glowbar: 1 - 40 mm

Nernst glower: 400 nm - 20 mm

Tungsten lamps are used in visible and near-infrared (NIR) absorption spectroscopy and the glowbar and Nernst glower are used for infrared spectroscopy.

DISCHARGE LAMPS

Discharge lamps, such as neon signs, pass an electric current through a rare gas or metal vapour to produce light. The electrons collide with gas atoms, exciting them to higher energy levels which then decay to lower levels by emitting light. Low-pressure lamps have sharp line emission characteristic of the atoms in the lamp, and high-pressure lamps have broadened lines superimposed on a continuum.

Common discharge lamps and their wavelength ranges are:

hydrogen or deuterium: 160 - 360 nm

mercury: 253.7 nm, and weaker lines in the near-uv and visible

Ne, Ar, Kr, Xe discharge lamps: many sharp lines throughout the near-uv to near-IR

xenon arc:·300 - 1300 nm

Deuterium lamps are the uv source in uv-vis absorption spectrophotometers. The sharp lines of the mercury and rare gas discharge lamps are useful for wavelength calibration of optical instrumentation. Mercury and xenon arc lamps are used to excite fluorescence.

HOLLOW-CATHODE LAMPS

Hollow-cathode lamps are a type of discharge lamp that produce narrow emission from atomic species. They get their name from the cup-shaped cathode, which is made from the element of interest. The electric discharge ionizes rare gas atoms, which are accelerated into the cathode and sputter metal atoms into the gas phase. Collisions with gas atoms or electrons excite the metal atoms to higher energy levels, which decay to lower levels by emitting light.

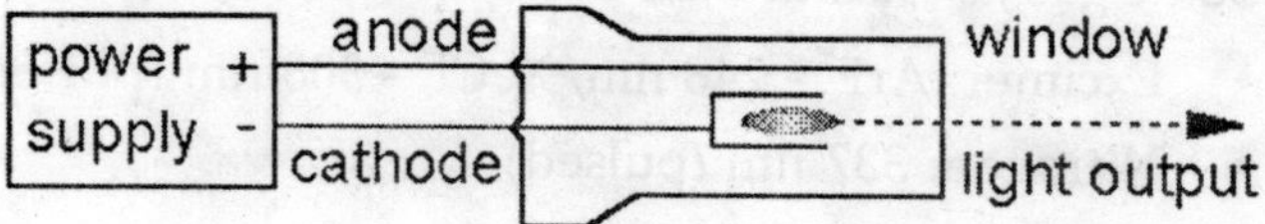

Fig. Schematic of a hollow-cathode lamp

Hollow-cathode lamps have become the most common light source for atomic absorption (AA) spectroscopy. They are also sometimes used as an excitation source for atomic-fluorescence spectroscopy (AFS).

LASERS

A laser is a coherent and highly directional radiation source. LASER stands for Light Amplification by Stimulated Emission of Radiation.

A laser consists of at least three components:

1. A gain medium that can amplify light that passes through it
2. An energy pump source to create a population inversion in the gain medium
3. Two mirrors that form a resonator cavity

The gain medium can be solid, liquid, or gas and the pump source can be an electrical discharge, a flashlamp, or another laser. The specific components of a laser vary depending on the gain medium and whether the laser is operated continuously (cw) or pulsed. The following headings describe specific laser designs.

GAS LASERS

Gas lasers are typically excited by an electrical discharge.

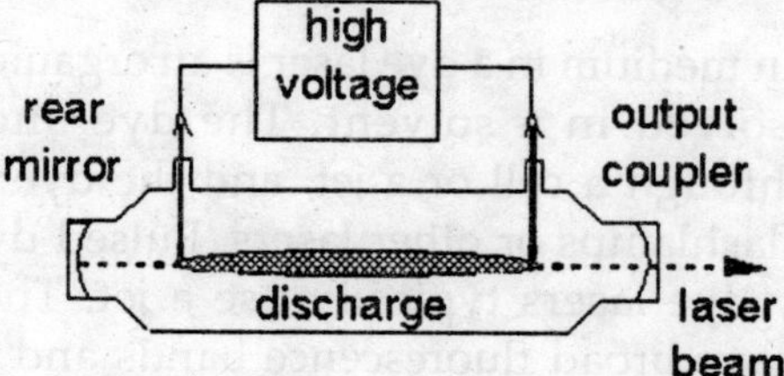

Fig. Schematic of a cw gas laser

Some gas lasers and their dominant lasing wavelength(s):

- Excimer: ArF^* - 248 nm, $XeCl^*$ - 308 nm (pulsed)
- Nitrogen: 337 nm (pulsed)
- He-Ne: 632.8 nm (cw)
- Ar ion: 488, 541 nm (cw)
- CO2: 10.6 μm (cw or pulsed)

SOLID-STATE LASERS

The gain medium in a solid-state laser is an impurity center in a crystal or glass. Solid-state lasers made from semiconductors are described below. The first laser was a ruby crystal (Cr^{3+} in Al_2O_3) that lased at 694 nm when pumped by a flashlamp. The most commonly used solid-state laser is one with Nd^{3+} in a $Y_3Al_5O_8$ (YAG) or $YLiF_4$(YLF) crystal or in a glass. These Nd^{3+} lasers operate either pulsed or cw and lase at approximately 1064 nm. The high energies of pulsed Nd^{3+}:YAG lasers allow efficient frequency doubling (532 nm), tripling (355 nm), or quadrupling (266 nm), and the 532 nm and 355 nm beams are commonly used to pump tunable dye lasers.

DYE LASERS

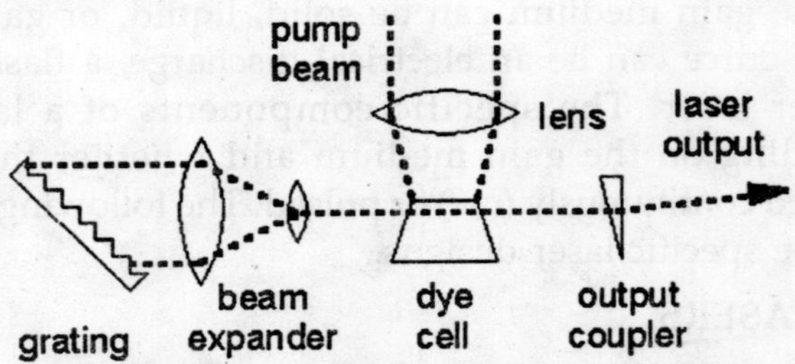

Fig. Schematic of a pulsed dye laser

The gain medium in a dye laser is an organic dye molecule that is dissolved in a solvent. The dye and solvent are circulated through a cell or a jet, and the dye molecules are excited by flashlamps or other lasers. Pulsed dye lasers use a cell and cw dye lasers typically use a jet. The organic dye molecules have broad fluorescence bands and dye lasers are typically tunable over 30 to 80 nm. Dyes exist to cover the near-uv to near-infrared spectral region: 330 - 1020 nm.

SEMICONDUCTOR LASERS

Semiconductor lasers are light-emitting diodes within a resonator cavity that is formed either on the surfaces of the diode or externally. An electric current passing through the diode produces light emission when electrons and holes recombine at the p-n junction. Because of the small size of the active medium, the laser output is very divergent and requires special optics to produce a good beam shape. These lasers are used in optical-fiber communications, CD players, and in high-resolution molecular spectroscopy in the near-infrared. Diode laser arrays can replace flashlamps to efficiently pump solid-state lasers. Diode lasers are tunable over a narrow range and different semiconductor materials are used to make lasers at 680, 800, 1300, and 1500 nm.

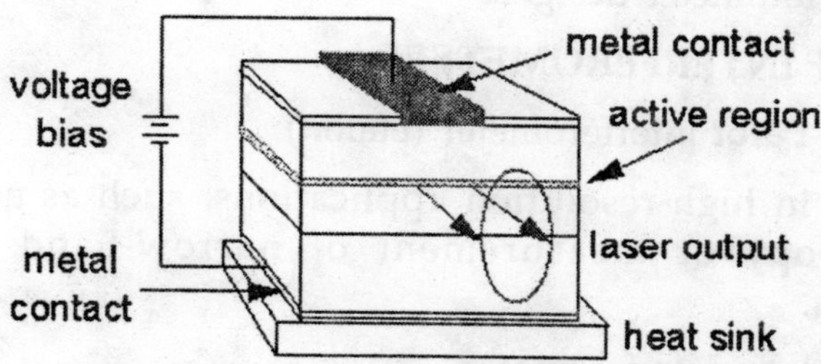

Fig. Schematic of a semiconductor diode laser

FILTERS

Filters separate different parts of the electromagnetic spectrum by absorbing or reflecting certain wavelengths and transmitting other wavelengths.

COLOUR FILTERS

Colour filters are glass substrates containing absorbing species that absorb certain wavelengths. A typical example is a cut-on colour filter, which blocks short wavelength light such as an excitation source, and transmits longer wavelength light such as fluorescence that reaches a detector.

INTERFERENCE FILTERS

Interference filters are made of multiple dielectric thin films on a substrate. They use interference to selectively transmit or reflect a certain range of wavelengths. A typical example is a bandpass interference filter that transmits a narrow range of wavelengths, and can isolate a single emission line from a discharge lamp.

INTERFEROMETERS

The purpose of an interferometer is similar to that of a filter or monochromator, i.e., to isolate a specific portion of the electromagnetic spectrum. Unlike prism or grating monochromators, interferometers are not dispersive instruments, but use interference to selectively transmit a certain wavelength. The links below lead to descriptions of three interferometer designs.

TYPES OF INTERFEROMETERS

Fabry-Perot interferometer (etalon)

Used in high-resolution applications, such as atomic spectroscopy or measurement of narrow-band laser linewidths.

Michelson interferometer

Used in fourier-transform infrared absorption spectrometers (FTIR).

Mach-Zender interferometer

Used to measure refractive index changes in gases and in interference microscopes to image transparent samples.

FABRY-PEROT INTERFEROMETER

The Fabry-Perot interferometer is commonly used as a narrow-bandpass filter or as an instrument to measure spectral linewidths.

Fabry-Perot interferometers that cannot be scanned are called etalons.

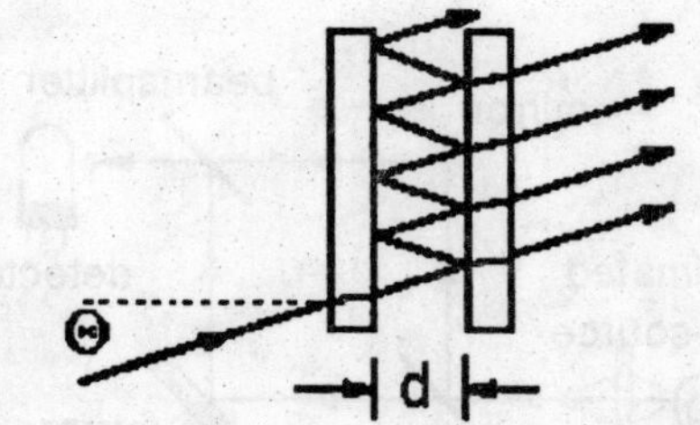

Fig. Schematic of a Fabry-Perot etalon

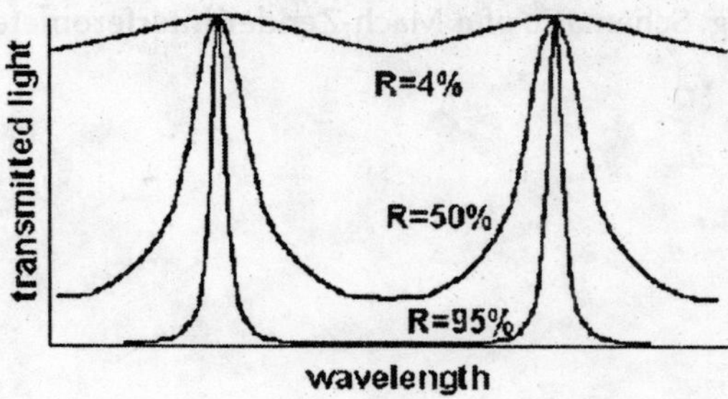

Fig. Transmission through a Fabry-Perot interferometer as a function of wavelength

MICHELSON INTERFEROMETER

The Michelson interferometer design is used in Fourier-transform infrared absorption spectrometers (FTIR).

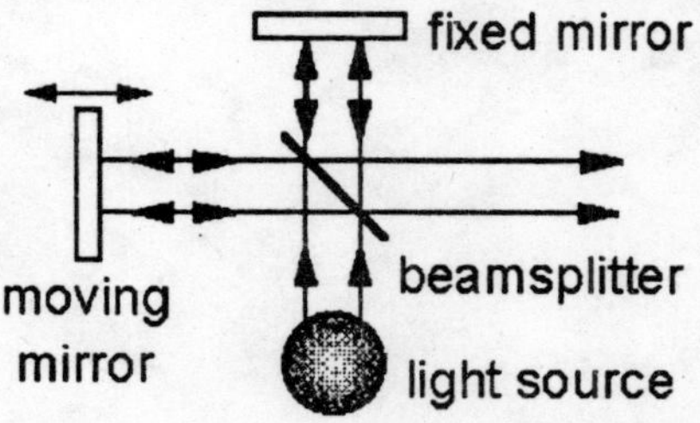

Fig. Schematic of a Michelson interferometer

MACH-ZENDER INTERFEROMETER

The Mach-Zender interferometer design is used to measure refractive index changes in gases, and in interference microscopes to image transparent samples.

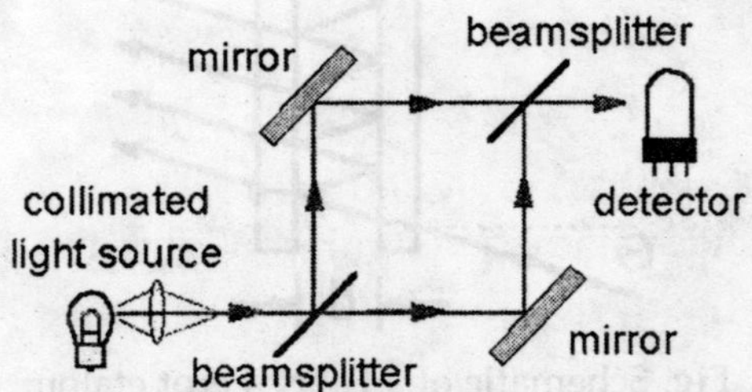

Fig. Schematic of a Mach-Zender interferometer

Chapter 17

Optical Radiation Detectors

Detectors convert light energy to an electrical signal. In spectroscopy, they are typically placed after a wavelength separator to detect a selected wavelength of light. Different types of detectors are sensitive in different parts of the electromagnetic spectrum.

Index of radiation detectors

- Charge-coupled detector array (CCD)
- Semiconductors (photodiodes and photovoltaics)
- Photodiode array (PDA)
- Photomultiplier tube (PMT)

CHARGE-COUPLED DEVICES (CCD)

A CCD is an integrated-circuit chip that contains an array of capacitors that store charge when light creates e-hole pairs. The charge accumulates and is read in a fixed time interval. CCDs are used in similar applications to other array detectors such as photodiode arrays, although the CCD is much more sensitive for measurement of low light levels.

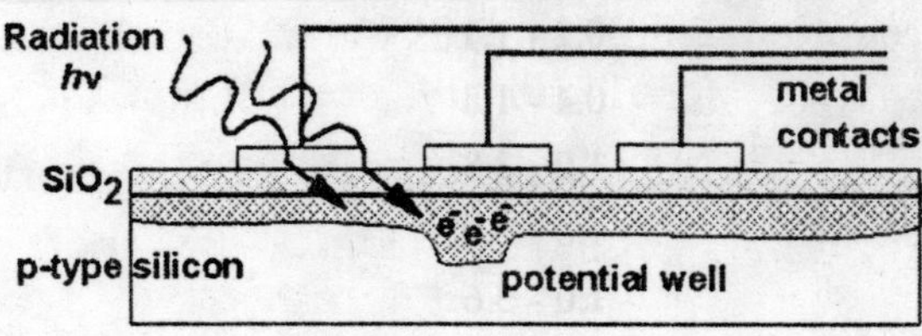

Fig. Schematic of a CCD

PHOTODIODE AND PHOTOVOLTAIC DETECTORS

When a photon strikes a semiconductor, it can promote an electron from the valence band (filled orbitals) to the conduction band (unfilled orbitals) creating an electron(-) - hole(+) pair. The concentration of these electron-hole pairs is dependent on the amount of light striking the semiconductor, making the semiconductor suitable as an optical detector. There are two ways to monitor the concentration of electron-hole pairs. In photodiodes, a voltage bias is present and the concentration of light-induced electron-hole pairs determines the current through semiconductor. Photovoltaic detectors contain a p-n junction, that causes the electron-hole pairs to separate to produce a voltage that can be measured.

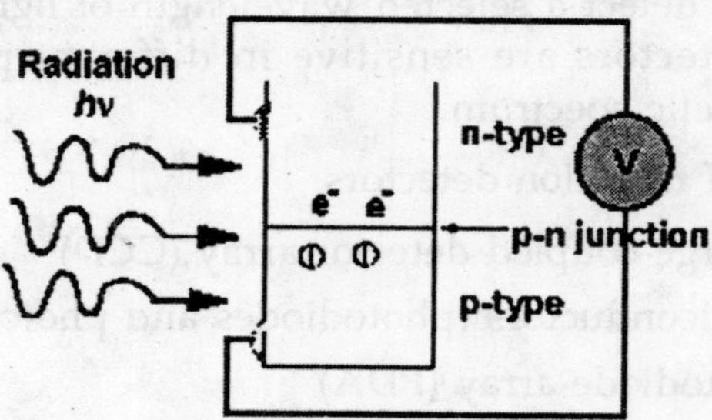

Fig. Schematic of semiconductor detector

Photodiode detectors are not as sensitive as PMTs but they are small and robust.

WAVELENGTH RANGE

Wavelength range

Detector type	λ (μ)
Si	0.2 - 1.1
Ge	0.4 - 1.8
InAs	1.0 - 3.8
InSb	1.0 - 7.0
InSb (77K)	1.0 - 5.6
HgCdTe (77K)	1.0 -25.0

PHOTODIODE ARRAY DETECTORS (PDA)

A photodiode array (PDA) is a linear array of discrete photodiodes on an integrated circuit (IC) chip. For spectroscopy it is placed at the image plane of a spectrometer to allow a range of wavelengths to be detected simultaneously. In this regard it can be thought of as an electronic version of photographic film. Array detectors are especially useful for recording the full uv-vis absorption spectra of samples that are rapidly passing through a sample flow cell, such as in an HPLC detector.

PDAs work on the same principle as simple photovoltaic detectors.

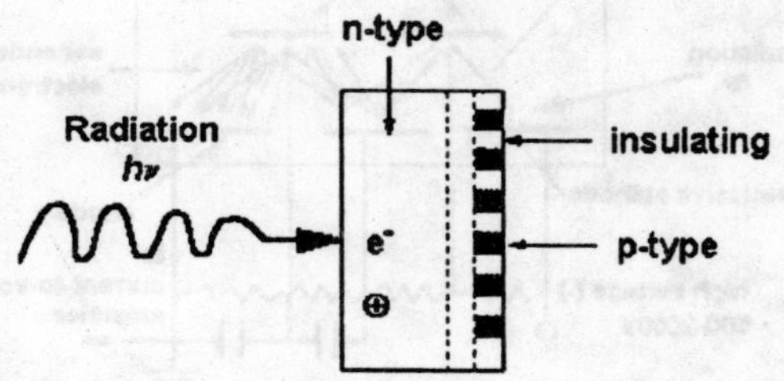

Fig. Schematic of a PDA

Light creates electron-hole pairs and the electrons migrate to the nearest PIN junction. After a fixed integration time the charge at each element is sequentially read with solid-state circuitry to generate the detector response as a function of linear distance along the array. PDAs are available with 512, 1024, or 2048 elements with typical dimensions of ~ 25 μm wide and 1-2 mm high.

PHOTOMULTIPLIER TUBE (PMT)

Photomultiplier Tubes (PMTS) are light detectors that are useful in low intensity applications such as fluorescence spectroscopy. Due to high internal gain, PMTs are very sensitive detectors.

PMTs are similar to phototubes. They consist of a photocathode and a series of dynodes in an evacuated glass enclosure. Photons that strikes the photoemissive cathode emits electrons due to the photoelectric effect. Instead of collecting these few electrons (there should not be a lot, since the primarily use for PMT is for verly low signal) at an anode like in the phototubes, the electrons are accelerated towards a series of additional electrodes called dynodes. These electrodes are each maintained at a more positive potential. Additional electrons are generated at each dynode. This cascading effect creates 10^5 to 10^7 electrons for each photon hitting the first cathode depending on the number of dynodes and the accelerating voltage. This amplified signal is finally collected at the anode where it can be measured.

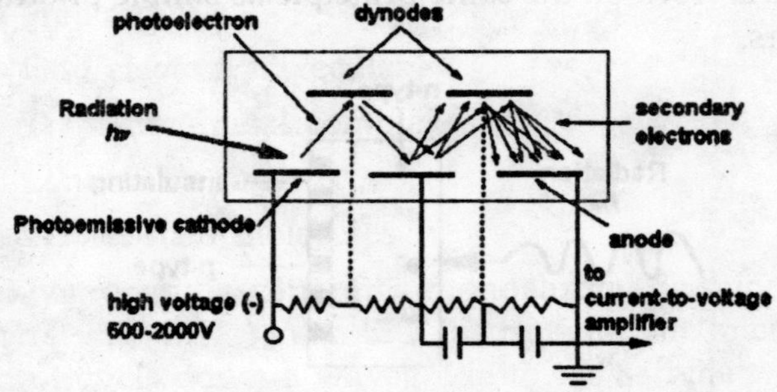

Fig. Schematic of a PMT

TYPICAL SPECIFICATIONS

Wavelength range: 110-1100 nm

(wavelength sensitivity dependent on wavelength, uv-sensitive PMTs must have uv-transmitting windows, see optical materials)

Chapter 18

Diffraction

INTRODUCTION

Diffraction is a wave property of electromagnetic radiation that causes the radiation to bend as it passes by an edge or through an aperture. Diffraction effects increase as the physical dimension of the aperture approaches the wavelength of the radiation. Diffraction of radiation results in interference that produces dark and bright rings, lines, or spots, depending on the geometry of the object causing the diffraction. Common interference effects for visible light are the rainbow pattern produced by an oil film on wet pavement and the diffraction of light from a narrow slit or a diffraction grating.

DIFFRACTION METHODS

A certain wavelength of radiation will constructively interfere when partially reflected between surfaces that produce a path difference equal to an integral number of wavelengths. This condition is described by the Bragg law:

$$n\lambda = 2d\sin\theta$$

where n is an integer, lambda is the wavelength of the radiation, d is the spacing between surfaces, and theta is the angle between the radiation and the surfaces. This relation demonstrates that interference effects are observable only when radiation interacts with physical dimensions that are approximately the same size as the wavelength of the radiation.

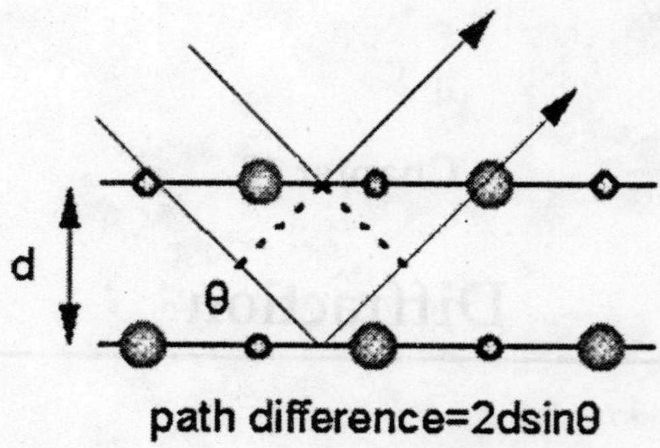

Fig. Interference of radiation between atomic planes in a crystal

These interference effects are useful for determining dimensions in solid materials, and therefore crystal structures. Since the distances between atoms or ions is on the order of 10^{-10} m (1 Å), diffraction methods require radiation in the x-ray region of the electromagnetic spectrum, or beams of electrons or neutrons with a similar wavelength. Electrons and neutrons are commonly thought of as particles, but they have wave properties with the wavelength depending on the energy of the particles as described by the de Broglie equation. The three diffraction methods have different properties that are described in more detail in separate documents. For example, the penetration depths of the three types of beams are quite different:

neutrons > x-rays > electrons.

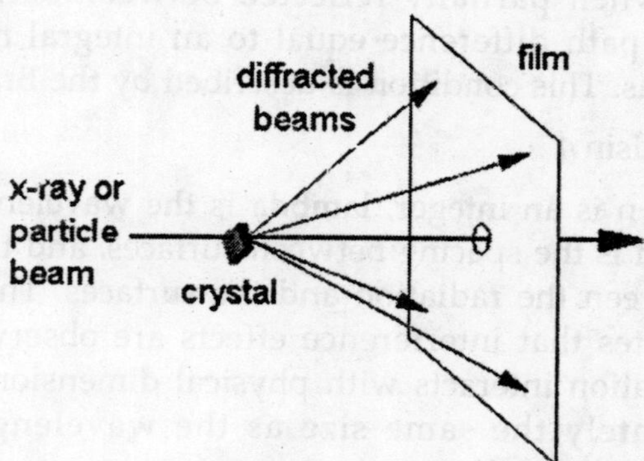

Fig. Schematic of crystal-structure determination by diffraction

ELECTRON DIFFRACTION

Electron diffraction provides similar structural information as neutron diffraction. Electron beams strongly interact with nuclei and electron diffraction is more useful than x-ray diffraction for determining proton positions. The strong interaction of electrons with matter results in a low penetration depth, and electron diffraction is usually used in a reflection geometry to study surfaces or thin films. Electron beams are easy to manipulate, detect, and focus to small spots to provide high spatial resolution.

Electrons scatter from gases and electron diffraction must be performed under vacuum.

NEUTRON DIFFRACTION

Neutron diffraction provides similar structural information as electron diffraction. Neutron beams interact more strongly with nuclei than do x-rays and neutron diffraction is more useful than x-ray diffraction for determining proton positions. Neutrons interact with a solid to a much lesser degree than x-rays and therefore have advantages in studying materials that are damaged by x-rays and in cases where a large penetration depth is desired. For the three types of diffraction methods, neutrons are unique in that they have a magnetic moment and are therefore sensitive to magnetic ordering in a solid.

Neutrons are produced by nuclear reactions in either a nuclear reactor or in an accelerator. Reactor sources produce a continuous spectrum of neutron energies and require a monochromator crystal to select a particular energy. Accelerator sources are usually operated in a pulsed mode and neutron wavelength is selected by time-of-flight methods, that is, data is taken at a fixed Bragg angle as a function of neutron energy.

X-RAY DIFFRACTION (XRD)

The wavelengths of x-rays are of the same order of magnitude as the distances between atoms or ions in a molecule or crystal (Å, 10^{-10} m). A crystal diffracts an x-ray beam passing through it to produce beams at specific angles depending on the x-ray wavelength, the crystal orientation, and the structure of the crystal. X-rays are predominantly diffracted by electron density and analysis of the diffraction angles produces an electron density map of the crystal. Since hydrogen atoms have very little electron density, determining their positions requires extensive refinement of the diffraction pattern. Electron diffraction and neutron diffraction are sensitive to nuclei and are often used to accurately determine hydrogen positions.

X-ray diffractometers consist of an x-ray generator, a goniometer and sample holder, and an x-ray detector such as photographic film or a movable proportional counter. X-ray tubes generate x-rays by bombarding a metal target with high-energy (10 - 100 keV) electrons that knock out core electrons. An electron in an outer shell fills thehole in the inner shell and emits an x-ray photon. Two common targets are Mo and Cu, which have strong K(alpha) x-ray emission at 0.71073 and 1.5418 Å, respectively. X-rays can also be generated by decelerating electrons in a target or a synchrotron ring. These sources produce a continuous spectrum of x-rays and require a crystal monochromator to select a single wavelength.

Fig. Picture of a single-crystal X-ray diffractometer

POWDER X-RAY DIFFRACTION

Powders of crytalline materials diffract x-rays. A beam of x-rays passing through a sample of randomly-oriented microcrystals produces a pattern of rings on a distant screen. Powder x-ray diffraction provides less information than single-crystal diffraction, however, it is much simpler and faster. Powder x-ray diffraction is useful for confirming the identity of a solid material and determining crystallinity and phase purity. Modern powder x-ray diffractometers consist of an x-ray source, a movable sample platform, an x-ray detector, and associated computer-controlled electronics. The sample is either packed into a shallow cup-shaped holder or deposited as a slurry onto a quatz substrate, and the sample holder spins slowly during the experiment to reduce sample heating. The x-ray source is usually the same as used in single-crystal diffractometers, Mo or Cu. The x-ray beam is fixed and the sample platform rotates with respect to the beam by an angle theta. The detector rotates at twice the rate of the sample and is at an angle of 2theta with respect to the incoming x-ray beame.

Chapter 19

Wave Optics

INTRODUCTION

Geometric optics is an incredibly successful theory. Probably its most important application is in describing and explaining the operation of commonly occurring optical instruments: *e.g.*, the camera, the telescope, and the microscope. Although geometric optics does not make any explicit assumption about the nature of light, it tends to suggest that light consists of a stream of massless particles. This is certainly what scientists, including, most notably, Isaac Newton, generally assumed up until about the year 1800.

Let us examine how the particle theory of light accounts for the three basic laws of geometric optics:

- *The law of geometric propagation:* This is easy. Massless particles obviously move in straight-lines in free space.
- *The law of reflection:* This is also fairly easy. We merely have to assume that light particles bounce *elastically* (*i.e.*, without energy loss) off reflecting surfaces.
- *The law of refraction:* This is the tricky one. Let us assume that the speed of light particles propagating through a transparent dielectric medium is proportional to the index of refraction,. Let us further assume that at a general interface between two different dielectric media, light particles crossing the

interface conserve momentum in the plane parallel to the interface. In general, this implies that the particle momenta normal to the interface are not conserved: *i.e.*, the interface exerts a normal reaction force on crossing particles, but no parallel force. From Fig. , parallel momentum conservation for light particles crossing the interface yields $n \equiv \sqrt{k}$.

$$\upsilon_1 \sin\theta_1 = \upsilon_2 \sin\theta_2$$

- However, by assumption $\upsilon_1 = n_1 c$, and $\upsilon_2 = n_2 c$, so

$$n_1 \sin\theta_1 = n_2 \sin\theta_2$$

- This highly contrived (and incorrect) derivation of the law of refraction was first proposed by Descartes in 1637. Note that it depends crucially on the (incorrect) assumption that light travels *faster* in dense media (*e.g.*, glass) than in rarefied media (*e.g.*, water). This assumption appears very strange to us nowadays, but it seemed eminently reasonable to scientists in the 17th and 18th centuries. After all, they knew that sound travels faster in dense media (*e.g.*, water) than in rarefied media (*e.g.*, air).

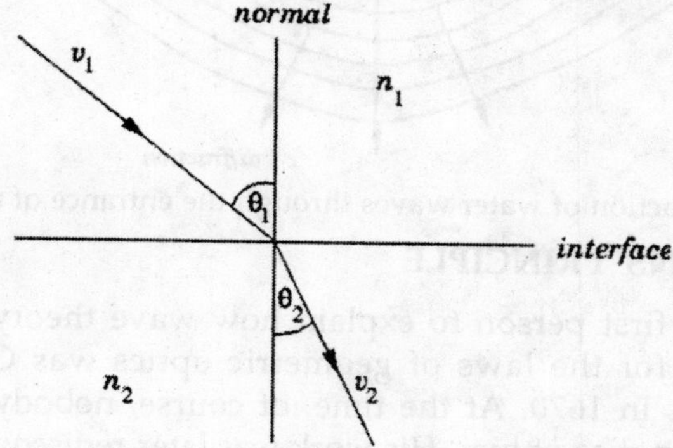

Fig. Descartes' model of refraction

The wave theory of light, which became established in the first half of the 19th century, initially encountered tremendous resistance. Let us briefly examine the reasons why scientists in the early 1800s refused to think of light as a wave phenomenon? Firstly, the particle theory of light was intimately associated with Isaac Newton, so any attack on this theory was considered to be a slight to his memory. Secondly, all of the waves that scientists were familiar with at that time manifestly did not travel in straight-lines. For instance, water waves are diffracted as they pass through the narrow mouth of a harbour, as shown in Fig.. In other words, the "rays" associated with such waves are bent as they traverse the harbour mouth. Scientists thought that if light were a wave phenomenon then it would also not travel in straight-lines: *i.e.*, it would not cast straight, sharp shadows, any more than water waves cast straight, sharp "shadows." Unfortunately, they did not appreciate that if the wavelength of light is much shorter than that of water waves then light can be a wave phenomenon and still propagate in a largely geometric manner.

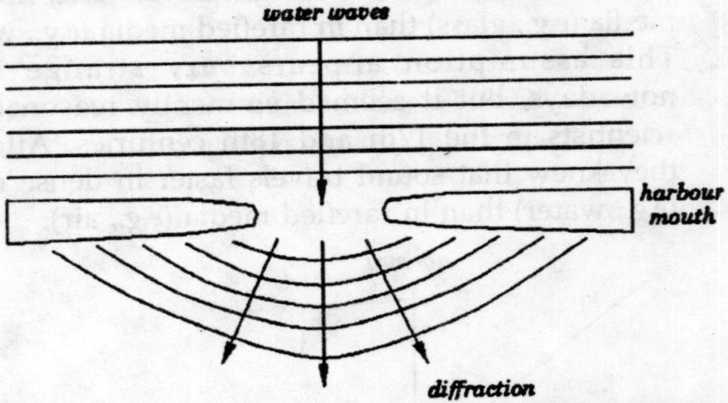

Fig. Refraction of water waves through the entrance of a harbour.

HUYGENS' PRINCIPLE

The first person to explain how wave theory can also account for the laws of geometric optics was Christiaan Huygens in 1670. At the time, of course, nobody took the slightest notice of him. His work was later rediscovered after the eventual triumph of wave theory.

Huygens had a very important insight into the nature of wave propagation which is nowadays called *Huygens' principle*. When applied to the propagation of light waves, this principle states that:

Every point on a wave-front may be considered a source of secondary spherical wavelets which spread out in the forward direction at the speed of light. The new wave-front is the tangential surface to all of these secondary wavelets.

According to Huygens' principle, a plane light wave propagates though free space at the speed of light,. The light rays associated with this wave-front propagate in straight-lines, as shown in Fig. It is also fairly straightforward to account for the laws of reflection and refraction using Huygens' principle.

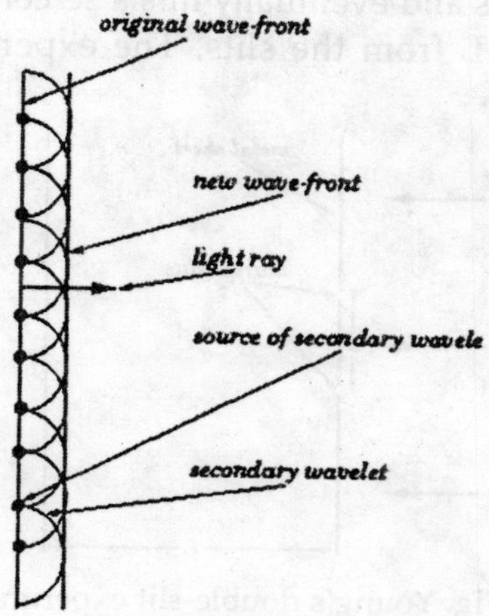

Fig. Huygen's principle.

YOUNG'S DOUBLE-SLIT EXPERIMENT

The first serious challenge to the particle theory of light was made by the English scientist Thomas Young in 1803. Young possessed one of the most brilliant minds in the history of science. A physician by training, he was the first to describe how the lens of the human eye changes shape in order to focus on objects at differing distances. He also studied Physics, and, amongst other things, definitely established the wave theory of light, as described below. Finally, he also studied Egyptology, and helped decipher the Rosetta Stone.

Young knew that sound was a wave phenomenon, and, hence, that if two sound waves of equal intensity, but 180°out of phase, reach the ear then they cancel one another out, and no sound is heard. This phenomenon is called *interference*. Young reasoned that if light were actually a wave phenomenon, as he suspected, then a similar interference effect should occur for light. This line of reasoning lead Young to perform an experiment which is nowadays referred to as *Young's double-slit experiment*.

In Young's experiment, two very narrow parallel slits, separated by a distance d, are cut into a thin sheet of metal. Monochromatic light, from a distant light-source, passes through the slits and eventually hits a screen a comparatively large distance L from the slits. The experimental setup is sketched in Fig. .

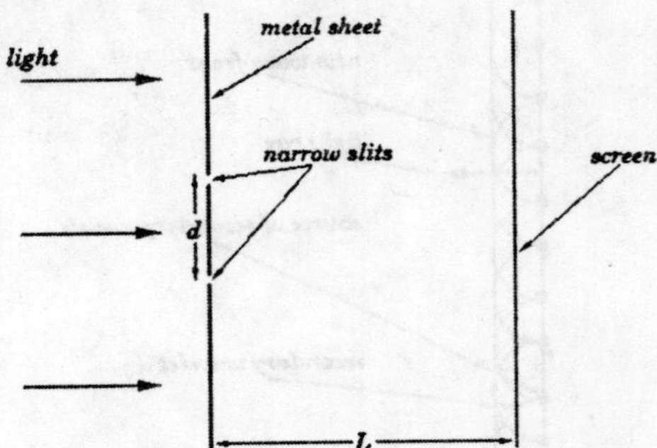

Fig. Young's double-slit experiment.

According to Huygens' principle, each slit radiates spherical light waves. The light waves emanating from each slit are superposed on the screen. If the waves are out of phase then *destructive interference* occurs, resulting in a dark patch on the screen. On the other hand, if the waves are completely in phase then *constructive interference* occurs, resulting in a light patch on the screen.

The point on the screen which lies exactly opposite to the centre point of the two slits, as shown in Fig. is obviously associated with a bright patch. This follows because the path-lengths from each slit to this point are the same. The waves emanating from each slit are initially in phase, since all points on the incident wave-front are in phase (*i.e.*, the wave-front is straight and parallel to the metal sheet). The waves are still in phase at point since they have traveled equal distances in order to reach that point.

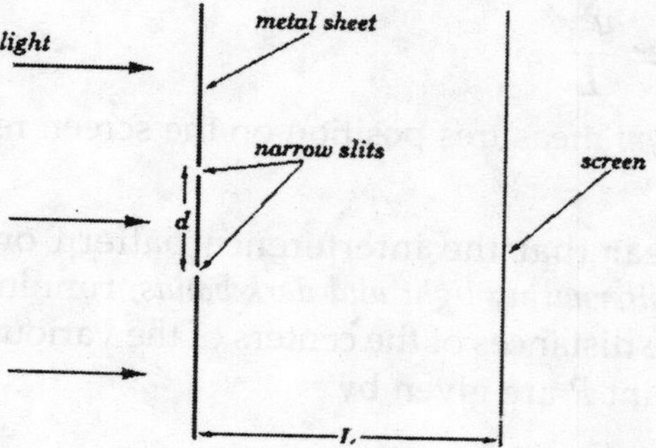

Fig. Interference of light in Young's double-slit experiment.

From the above discussion, the general condition for constructive interference on the screen is simply that the difference in path-length Δ between the two waves be an *integer* number of wavelengths. In other words,

$$\Delta = m\lambda$$

where m =0,1,2,.... Of course, the point corresponds to the special case where. It follows, from Fig. that the angular location of the th bright patch on the screen is given by

$$\sin\theta_m = \frac{\Delta}{d} = \frac{m\lambda}{d}.$$

Likewise, the general condition for destructive interference on the screen is that the difference in path-length between the two waves be a *half-integer* number of wavelengths. In other words,

$$\Delta = (m + 1/2)\,\lambda,$$

where m =0,1,2,.... It follows that the angular coordinate of the th dark patch on the screen is given by

$$\sin\theta'_m = \frac{\Delta}{d} = \frac{(m+1/2)\,\lambda}{d}.$$

Usually, we expect the wavelength λ of the incident light to be much less than the perpendicular distance L to the screen. Thus,

$$\sin\theta_m \simeq \frac{y_m}{L};$$

where ym measures position on the screen relative to the point P.

It is clear that the interference pattern on the screen consists of *alternating light and dark bands,* running parallel to the slits. The distances of the centers of the various light bands from the point P are given by

$$y_m = \frac{m\,\lambda\,L}{d};$$

where m = 1,2,3,... Likewise, the distances of the centres of the various dark bands from the point P are given by

$$y'_m = \frac{(m+1/2)\,\lambda\,L}{d};$$

where m = 1,2,3,... The bands are *equally spaced,* and of thickness $\lambda L/d$. Note that if the distance from the screen L is much larger than the spacing between the two slits then the thickness of the bands on the screen greatly exceeds the wavelength λ of the light. Thus, given a sufficiently large ratio L/d, it should be possible to observe a banded interference pattern on the screen, despite the fact that the wavelength of visible light is only of order 1 micron. Indeed, when Young performed this experiment in 1803 he observed an interference pattern of the type described above. Of course, this pattern is a *direct consequence* of the wave nature of light, and is completely inexplicable on the basis of geometric optics.

It is interesting to note that when Young first presented his findings to the Royal Society of London he was ridiculed. His work only achieved widespread acceptance when it was confirmed, and greatly extended, by the French physicists Augustin Fresnel and Francois Argo in the 1820s. The particle theory of light was dealt its final death-blow in 1849 when the

French physicists Fizeau and Foucault independently demonstrated that light propagates *more slowly* though water than though air. That the particle theory of light can only account for the law of refraction on the assumption that light propagates *faster* through dense media, such as water, than through rarefied media, such as air.

INTERFERENCE IN THIN FILMS

In everyday life, the interference of light most commonly gives rise to easily observable effects when light impinges on a thin film of some transparent material. For instance, the brilliant colours seen in soap bubbles, in oil films floating on puddles of water, and in the feathers of a peacock's tail, are due to interference of this type.

Suppose that a very thin film of air is trapped between two pieces of glass, as shown in Fig. If monochromatic light (*e.g.*, the yellow light from a sodium lamp) is incident *almost normally* to the film then some of the light is reflected from the interface between the bottom of the upper plate and the air, and some is reflected from the interface between the air and the top of the lower plate. The eye focuses these two parallel light beams at one spot on the retina. The two beams produce either destructive or constructive interference, depending on whether their path difference is equal to an odd or an even number of half-wavelengths, respectively.

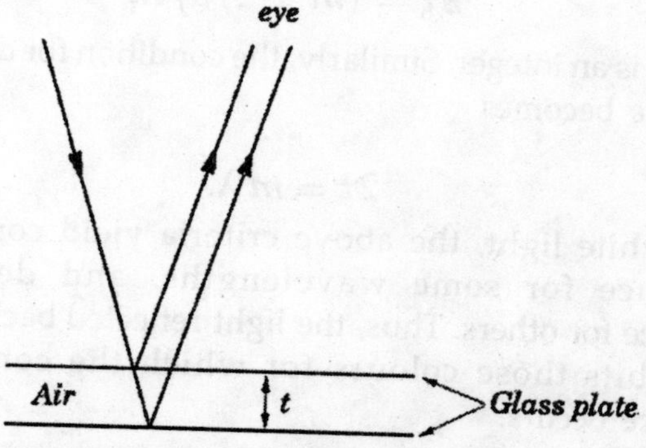

Fig. Interference of light due to a thin film of air trapped between two pieces of glass.

Let be the thickness of the air film. The difference in path-lengths between the two light rays shown in the figure is clearly $\Delta = 2t$. Naively, we might expect that constructive interference, and, hence, *brightness*, would occur if $\Delta = m\lambda$, where is an integer, and destructive interference, and, hence, *darkness*, would occur if $\Delta = (m+1/2)\lambda$. However, this is not the entire picture, since an additional phase difference is introduced between the two rays on reflection. The first ray is reflected at an interface between an optically dense medium (glass), through which the ray travels, and a less dense medium (air). There is no phase change on reflection from such an interface, just as there is no phase change when a wave on a string is reflected from a free end of the string. (Both waves on strings and electromagnetic waves are *transverse waves*, and, therefore, have analogous properties.) The second ray is reflected at an interface between an optically less dense medium (air), through which the ray travels, and a dense medium (glass). There is a 180° phase change on reflection from such an interface, just as there is a 180° phase change when a wave on a string is reflected from a fixed end. Thus, an additional 180° phase change is introduced between the two rays, which is equivalent to an additional path difference of λ/2. When this additional phase change is taken into account, the condition for constructive interference becomes

$$2t = (m + 1/2)\,\lambda,$$

where is an integer. Similarly, the condition for destructive interference becomes

$$2t = m\,\lambda.$$

For white light, the above criteria yield constructive interference for some wavelengths, and destructive interference for others. Thus, the light reflected back from the film exhibits those colours for which the constructive interference occurs.

If the thin film consists of water, oil, or some other transparent material of refractive index then the results are basically the same as those for an air film, except that the wavelength of the light in the film is reduced from λ (the vacuum wavelength) to λ/n. It follows that the modified criteria for constructive and destructive interference are

$$2nt = (m + 1/2)\,\lambda;$$

and

$$2nt = m\,\lambda;$$

respectively.

Chapter 20

Nuclear Physics

FISSION AND FUSION

Fission of radioactive elements was already well established in the early part of the century, and activation by neutrons, to generate more unstable isotopes, was investigated before fission of natural isotopes was seen. The inverse process, fusion, was understood somewhat later, and Niels Bohr developped a model describing the nucleus as a fluid drop. This model - the collective model - was further developped by his son Aage Bohr and Ben Mottelson. A very different model of the nucleus, the shell model, was designed by Maria Goeppert-Mayer and Hans Jensen in 1952, concentrating on individual nucleons. The dichotomy between a description as individual particles and as a collective whole characterises much of "low-energy" nuclear physics.

LOW-ENERGY NUCLEAR PHYSICS

The field of low-energy nuclear physics, which concentrates mainly on structure of and low-energy reaction on nuclei, has become one of the smaller parts of nuclear physics (apart from in the UK). Notable results have included better understanding of the nuclear medium, high-spin physics, superdeformation and halo nuclei. Current experimental interest is in those nuclei near the "driplines" which are of astrophysical importance, as well as of other interest.

MEDIUM-ENERGY NUCLEAR PHYSICS

Medium energy nuclear physics is interested in the response of a nucleus to probes at such energies that we can no longer consider nucleons to be elementary particles. Most modern experiments are done by electron scattering, and concentrate on the role of QCD (see below) in nuclei, the structure of mesons in nuclei and other complicated questions.

HIGH-ENERGY NUCLEAR PHYSICS

This is not a very well-defined field, since particle physicists are also working here. It is mainly concerned with ultra-relativistic scattering of nuclei from each other, addressing questions about the quark-gluon plasma. It should be nuclear physics, since we consider "dirty" systems of many particles, which are what nuclear physicists are good at.

MESONS, LEPTONS AND NEUTRINOS

In 1934 Yukawa introduces a new particle, the pion (π), which can be used to describe nuclear binding. He estimates it's mass at 200 electron masses. In 1937 such a particle is first seen in cosmic rays. It is later realised that it interacts too weakly to be the pion and is actually a lepton (electron-like particle) called the μ. The is found (in cosmic rays) and is the progenitor of the $^{\mu}$'s that were seen before:

$$\pi^+ \to \mu^+ + \nu_\mu$$

The next year artificial pions are produced in an accelerator, and in 1950 the neutral pion is found,

$$\pi^0 \to \gamma\gamma .$$

This is an example of the conservation of electric charge. Already in 1938 Stuckelberg had found that there are other conserved quantities: the number of baryons (*n* and *p* and...) is also conserved!

After a serious break in the work during the latter part of WWII, activity resumed again. The theory of electrons and positrons interacting through the electromagnetic field (photons) was tackled seriously, and with important contributions of (amongst others) Tomonaga, Schwinger and Feynman was developed into a highly accurate tool to describe hyperfine structure.

Experimental activity also resumed. Cosmic rays still provided an important source of extremely energetic particles, and in 1947 a "strange" particle (K^+ was discovered through its very peculiar decay pattern. Balloon experiments led to additional discoveries: So-called V particles were found, which were neutral particles, identified as the Λ^0 and K^0. It was realised that a new conserved quantity had been found. It was called strangeness.

The technological development around WWII led to an explosion in the use of accelerators, and more and more particles were found. A few of the important ones are the antiproton, which was first secn in 1955, and the Δ, a very peculiar excited state of the nucleon, that comes in four charge states $\Delta^{++}, \Delta^{+}, \Delta^{0}, \Delta^{-}$.

Theory was develop-ping rapidly as well. A few highlights: In 1954 Yang and Mills develop the concept of gauged Yang-Mills fields. It looked like a mathematical game at the time, but it proved to be the key tool in developing what is now called "the standard model".

In 1956 Yang and Lee make the revolutionary suggestion that parity is not necessarily conserved in the weak interactions. In the same year "madam" CS Wu and Alder show experimentally that this is true: God is weakly left-handed!

In 1957 Schwinger, Bludman and Glashow suggest that all weak interactions (radioactive decay) are mediated by the charged bosons $W^{\pm}$. In 1961 Gell-Mann and Ne'eman introduce the "eightfold way": a mathematical taxonomy to organize the particle zoo.

THE SUB-STRUCTURE OF THE NUCLEON (QCD)

In 1964 Gell-mann and Zweig introduce the idea of quarks: particles with spin 1/2 and fractional charges. They are called up, down and strange and have charges 2/3, - 1/3, - 1/3 times the electron charge.

Since it was found (in 1962) that electrons and muons are each accompanied by their own neutrino, it is proposed to organize the quarks in multiplets as well:

$e\ \nu_e$ (u, d)

$\mu \nu_\mu$ (s, c)

This requires a fourth quark, which is called charm.

In 1965 Greenberg, Han and Nambu explain why we can't see quarks: quarks carry colour charge, and all observe particles must have colour charge 0. Mesons have a quark and an antiquark, and baryons must be build from three quarks through its peculiar symmetry.

The first evidence of quarks is found (1969) in an experiment at SLAC, where small pips inside the proton are seen. This gives additional impetus to develop a theory that incorporates some of the ideas already found: this is called QCD. It is shown that even though quarks and gluons (the building blocks of the theory) exist, they cannot be created as free particles. At very high energies (very short distances) it is found that they behave more and more like real free particles. This explains the SLAC experiment, and is called asymptotic freedom.

The J/ψ meson is discovered in 1974, and proves to be the $c\bar{c}$ bound state. Other mesons are discovered (D0, $\overline{\mu}\ c$) and agree with QCD.

In 1976 a third lepton, a heavy electron, is discovered (τ). This was unexpected! A matching quark (b for bottom or beauty) is found in 1977. Where is its partner, the top? It will only be found in 1995, and has a mass of 175 GeV/c^2 (similar to a lead nucleus...)! Together with the conclusion that there are no further light neutrinos (and one might hope no quarks and charged leptons) this closes a chapter in particle physics.

THE $W^{\pm}$AND Z BOSONS

On the other side a electro-weak interaction is developed by Weinberg and Salam. A few years later 't Hooft shows that it is a well-posed theory. This predicts the existence of three extremely heavy bosons that mediate the weak force: the Z^0 and the $W^{\pm}$. These have been found in 1983. There is one more particle predicted by these theories: the Higgs particle. Must be very heavy!

GUTS, SUPERSYMMETRY, SUPERGRAVITY

This is not the end of the story. The standard model is surprisingly inelegant, and contains way to many parameters for theorists to be happy. There is a dark mass problem in astrophysics - most of the mass in the universe is not seen! This all leads to the idea of an underlying theory. Many different ideas have been developed, but experiment will have the last word! It might already be getting some signals: researchers at DESY see a new signal in a region of particle that are 200 GeV heavy - it might be noise, but it could well be significant!

There are several ideas floating around: one is the grand-unified theory, where we try to comine all the disparate forces in nature in one big theoretical frame. Not unrelated is the idea of supersymmetries: For every "boson" we have a "fermion". There are some indications that such theories may actually be able to make useful predictions.

EXTRATERRESTRIAL PARTICLE PHYSICS

One of the problems is that it is difficult to see how e can actually build a microscope that can look a a small enough scale, i.e., how we can build an accelerator that will be able to accelarte particles to high enough energies? The answer is simple - and has been more or less the same through the years: Look at the cosmos. Processes on an astrophysical scale can have amazing energies.

BALLOON EXPERIMENTS

One of the most used techniques is to use balloons to send up some instrumentation. Once the atmosphere is no longer the perturbing factor it normally is, one can then try to detect interesting physics. A problem is the relatively limited payload that can be carried by a balloon.

GROUND BASED SYSTEMS

These days people concentrate on those rare, extremely high energy processes (of about 10^{29} eV), where the effect of the atmosphere actually help detection. The trick is to look at showers of (lower-energy) particles created when such a high-energy particle travels through the earth's atmosphere.

DARK MATTER

One of the interesting cosmological questions is whether we live in an open or closed universe. From various measurements we seem to get conflicting indications about the mass density of (parts of) the universe. It seems that the ration of luminous to non-luminous matter is rather small. Where is all that "dark mass": Mini-jupiters, small planetoids, dust, or new particles....

(SOLAR) NEUTRINOS

The neutrino is a very interesting particle. Even though we believe that we understand the nuclear physics of the sun, the number of neutrinos emitted from the sun seems to anomalously small. Unfortunately this is very hard to measure, and one needs quite a few different experiments to disentangle the physics behind these processes. Such experiments are coming on line in the next few years. These can also look at neutrinos coming from other astrophysical sources, such as supernovas, and enhance our understanding of those processes. Current indications from Kamiokande are that neutrinos do have mass, but oscillation problems still need to be resolved.

Chapter 21

Accelerators

RESOLVING POWER

Both nuclear and particle physics experiments are typically performed at accelerators, where particles are accelerated to extremely high energies, in most cases relativistic (i.e., $v \approx c$). To understand why this happens we need to look at the rôle the accelerators play. Accelerators are nothing but extremely big microscopes. At ultrarelativistic energies it doesn't really matter what the mass of the particle is, its energy only depends on the momentum:

$$E = h\nu = \approx pc$$

The typical resolving power of a microscope is about the size of one wave-length, λ. For an an ultrarelativistic particle this implies an energy of

$$E = pc = h\frac{c}{\lambda}$$

Size and energy-scale for various objects

particle	scale	energy
atom	$10^{-10}m$	2 keV
nucleus	$10^{-14}m$	20 MeV
nucleon	$10^{-15}m$	200 MeV
quark?	$< 10^{-18}m$	>200 GeV

You may not immediately appreciate the enormous scale of these energies. An energy of 1 TeV (= 10^{12}eV) is 3×10^{-7} J, which is the same as the kinetic energy of a 1g particle moving at 1.7 cm/s. And that for particles that are of submicroscopic size! We shall thus have to push these particles very hard indeed to gain such energies. In order to push these particles we need a handle to grasp hold of. The best one we know of is to use charged particles, since these can be accelerated with a combination of electric and magnetic fields - it is easy to get the necessary power as well.

TYPES

We can distinguish accelerators in two ways. One is whether the particles are accelerated along straight lines or along (approximate) circles. The other distinction is whether we used a DC (or slowly varying AC) voltage, or whether we use radio-frequency AC voltage, as is the case in most modern accelerators.

DC FIELDS

Acceleration in a DC field is rather straightforward: If we have two plates with a potential *V* between them, and release a particle near the plate at lower potential it will be accelerated to an energy $\frac{1}{2}mv^2 = eV$. This was the original technique that got Cockroft and Wolton their Nobel prize.

VAN DER GRAAFF GENERATOR

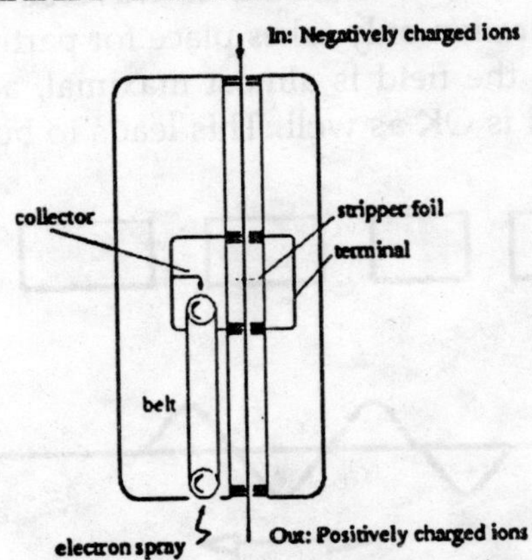

Fig. A sketch of a tandem van der Graaff generator

A better system is the tandem van der Graaff generator, even though this technique is slowly becoming obsolete in nuclear physics (technological applications are still very common). The idea is to use a (non-conducting) rubber belt to transfer charge to a collector in the middle of the machine, which can be used to build up sizeable (20 MV) potentials. By sending in negatively charged ions, which are stripped of (a large number of) their electrons in the middle of the machine we can use this potential twice. This is the mechanism used in part of the Daresbury machine.

OTHER LINEAR ACCELERATORS

Linear accelerators (called Linacs) are mainly used for electrons. The idea is to use a microwave or radio frequency field to accelerate the electrons through a number of connected cavities (DC fields of the desired strength are just impossible to maintain). A disadvantage of this system is that electrons can only be accelerated in tiny bunches, in small parts of the time. This so-called "duty-cycle", which is small (less than a percent) makes these machines not so beloved. It is also hard to use a linac in colliding beam mode (see below).

There are two basic setups for a linac. The original one is to use elements of different length with a fast oscillating (RF) field between the different elements, designed so that it takes exactly one period of the field to traverse each element. Matched acceleration only takes place for particles traversing the gaps when the field is almost maximal, actually sightly before maximal is OK as well. This leads to bunches coming out.

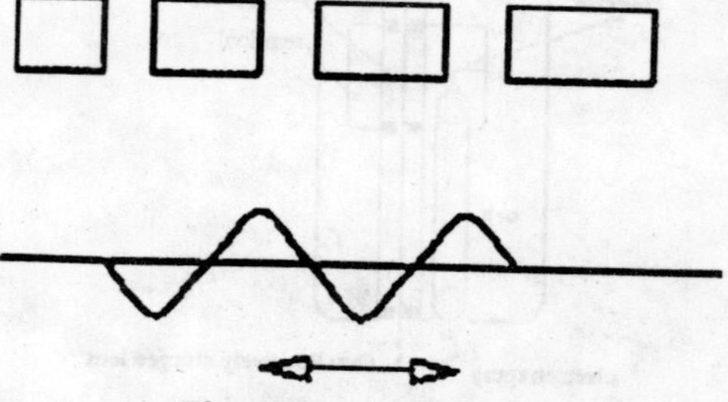

Fig. Figure of a linac

More modern electron accelerators are build using microwave cavities, where standing microwaves are generated. Such a standing wave can be thought of as one wave moving with the electron, and another moving the other wave. If we start of with relativistic electrons, $v \approx c$, this wave accelerates the electrons. This method requires less power than the one above.

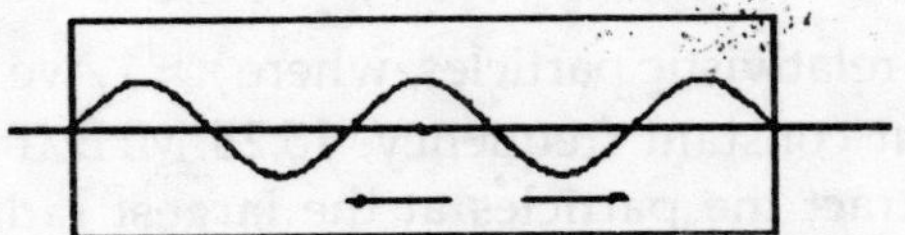

Fig. Acceleration by a standing wave

CYCLOTRON

The original design for a circular accelerator dates back to the 1930's, and is called a cyclotron. Like all circular accelerators it is based on the fact that a charged particle (charge qe) in a magnetic field B with velocity v moves in a circle of radius r, more precisely

$$\frac{\gamma m v^2}{r}$$

$qvB =$,

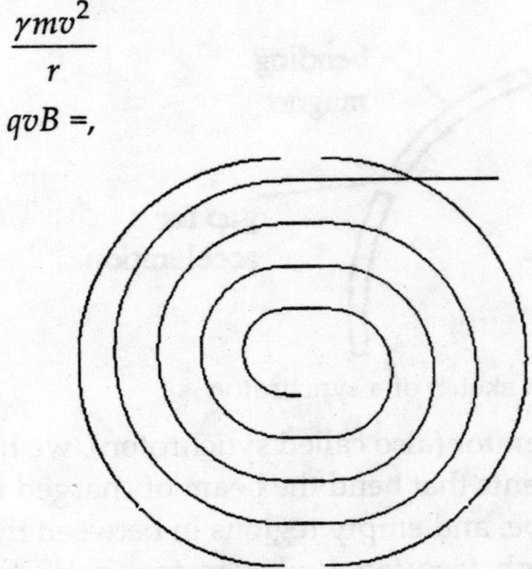

Fig. A sketch of a cyclotron

where γm is the relativistic mass, $\gamma = (1 - \beta^2)^{-1/2}$, $\beta = v/c$. A cyclotron consists of two metal "D"-rings, in which the particles are shielded from electric fields, and an electric field is applied between the two rings, changing sign for each half-revolution. This field then accelerates the particles.

The field has to change with a frequency equal to the angular velocity,

$$f = \frac{\omega}{2\pi} = \frac{\upsilon}{2\pi r} = \frac{qB}{2\pi\gamma m}$$

For non-relativistic particles, where $\gamma \approx 1$, we can thus run a cyclotron at constant frequency, 15.25 MHz/T for protons. Since we extract the particles at the largest radius possible, we can determine the velocity and thus the energy,

$$E = mc^2 = [(qBRc)^2 + m^2c^4]^{1/2}$$

SYNCHROTON

The shear size of a cyclotron that accelerates particles to 100 GeV or more would be outrageous. For that reason a different type of accelerator is used for higher energy, the so-called synchroton where the particles are accelerated in a circle of constant diameter.

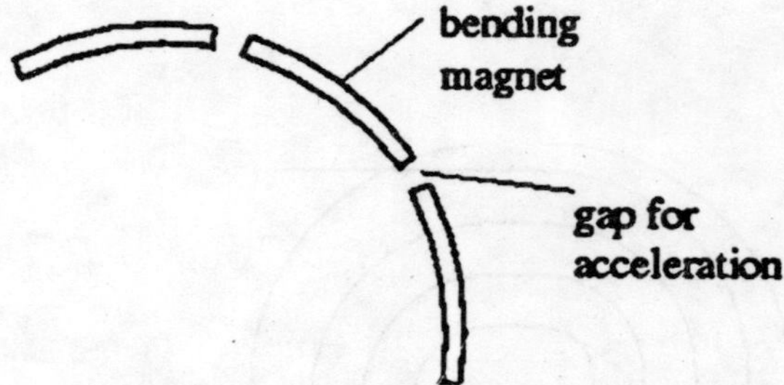

Fig. A sketch of a synchroton

In a circular accelerator (also called synchroton), we have a set of magnetic elements that bend the beam of charged into an almost circular shape, and empty regions in between those elements where a high frequency electro-magnetic field accelerates the particles to ever higher energies. The particles make many passes through the accelerator, at every increasing momentum. This makes critical timing requirements on the accelerating fields, they cannot remain constant.

Using the equations given above, we find that

$$f = \frac{qB}{2\pi\gamma m} = \frac{qBc^2}{2\pi E} = \frac{qBc^2}{2\pi(m^2c^4 + q^2B^2R^2c^2)^{1/2}}$$

For very high energy this goes over to

$$f = \frac{c}{2\pi R}, \qquad E = qBRc,$$

so we need to keep the frequency constant whilst increasing the magnetic field. In between the bending elements we insert (here and there) microwave cavities that accelerate the particles, which leads to bunching, i.e., particles travel with the top of the field.

So what determines the size of the ring and its maximal energy? There are two key factors:

As you know, a free particle does not move in a circle. It needs to be accelerated to do that. The magnetic elements take care of that, but an accelerated charge radiates - That is why there are synchroton lines at Daresbury! The amount of energy lost through radiation in one pass through the ring is given by (all quantities in SI units)

$$\Delta E = \frac{4\pi}{3\epsilon_0}\frac{q^2\beta^3\gamma^4}{R}$$

with $\beta = v/c$, $\gamma = 1/\sqrt{1-\beta^2}$ and R is the radius of the accelerator in meters. In most cases $v \approx c$, and we can replace by 1. We can also use one of the charges to re-express the energy-loss in eV:

$$\Delta_E \approx \frac{4\pi}{3\epsilon_0}\frac{q\gamma^4}{R}\Delta_E \approx \frac{4\pi}{3\epsilon_0}\frac{q}{R}\left(\frac{E}{mc^2}\right)^4$$

Thus the amount of energy lost is proportional to the fourth power of the relativistic energy, $E = \lambda\, mc^2$. For an electron at 1 TeV energy λ is

$$\gamma_e = \frac{E}{m_e c^2} = \frac{10^{12}}{511\times10^3} = 1.9\times10^6$$

and for a proton at the same energy

$$\gamma_p = \frac{E}{m_p c^2} = \frac{10^{12}}{939 \times 10^6} = 1.1{\times}10^3$$

This means that a proton looses a lot less energy than an electron (the fourth power in the expression shows the difference to be 10^{12}!). Let us take the radius of the ring to be 5 km (large, but not extremely so). We find the results listed in table.

Energy loss for a proton or electron in a synchroton of radius 5km

proton	E	ΔE
	1 GeV	$1.5{\times}10^{-11}$ eV
	10 GeV	$1.5{\times}10^{-7}$ eV
	100 GeV	$1.5{\times}10^{-3}$ eV
	1000 GeV	$1.5{\times}10^{1}$ eV
electron	E	ΔE
	1 GeV	$2.2{\times}10^{2}$ eV
	10 GeV	2.2 MeV
	100 GeV	22 GeV
	1000 GeV	$2.2{\times}10^{15}$ GeV

The other key factor is the maximal magnetic field. From the standard expression for the centrifugal force we find that the radius R for a relativistic particle is related to it's momentum (when expressed in GeV/c) by

$$p = 0.3BR$$

For a standard magnet the maximal field that can be reached is about 1T, for a superconducting one 5T. A particle moving at $p = 1TeV/c = 1000GeV/c$ requires a radius of

Table: Radius R of an synchroton for given magnetic fields and momenta.

B	p	R
1 T	1 GeV/c	3.3 m
	10 GeV/c	33 m

	100 GeV/c	330 m
	1000 GeV/c	3.3 km
5 T	1 GeV/c	0.66 m
	10 GeV/c	6.6 m
	100 GeV/c	66 m
	1000 GeV/c	660 m

TARGETS

There are two ways to make the necessary collisions with the accelerated beam: Fixed target and colliding beams.

In fixed target mode the accelerated beam hits a target which is fixed in the laboratory. Relativistic kinematics tells us that if a particle in the beam collides with a particle in the target, their centre-of-mass (four) momentum is conserved. The only energy remaining for the *reaction* is the relative energy (or energy within the cm frame). This can be expressed as

$$E_{CM} = [m_b{}^2c^4 + m_t{}^2c^4 + 2m_tc^2E_L]^{1/2}$$

where m_b is the mass of a beam particle, m_t is the mass of a target particle and E_L is the beam energy as measured in the laboratory. as we increase E_L we can ignore the first tow terms in the square root and we find that

$$E_{CM} \approx \sqrt{2mtc^2 EL}\,'$$

and thus the centre-of-mass energy only increases as the square root of the lab energy

In the case of colliding beams we use the fact that we have (say) an electron beam moving one way, and a positron beam going in the opposite direction. Since the centre of mass is at rest, we have the full energy of both beams available,

$$E_{CM} = 2E_L.$$

This grows linearly with lab energy, so that a factor two increase in the beam energy also gives a factor two increase in the available energy to produce new particles! We would only have gained a factor $\sqrt{2}$ for the case of a fixed target. This is the reason that almost all modern facilities are colliding beams.

THE MAIN EXPERIMENTAL FACILITIES

Let me first list a couple of facilities with there energies, and then discuss the facilities one-by-one.

Table: Fixed target facilities, and their beam energies

accelerator	facility	particle	energy
KEK	Tokyo	p	12 GeV
SLAC	Stanford	e^-	25GeV
PS	CERN	p	28 GeV
AGS	BNL	p	32 GeV
SPS	CERN	p	250 GeV
Tevatron II	FNL	p	1000 GeV

Table: Colliding beam facilities, and their beam energies

accelerator	facility	particle and energy (in GeV)
CESR	Cornell	$e^+(6) + e^-(6)$
PEP	Stanford	$e^+(15) + e^-(15)$
Tristan	KEK	$e^+(32) + e^-(32)$
SLC	Stanford	$e^+(50) + e^-(50)$
LEP	CERN	$e^+(60) + e^-(60)$
SppS	CERN	$p(450) + \bar{p}\ (450)$
Tevatron I	FNL	$p(1000) + \bar{p}\ (1000)$
LHC	CERN	$e^-(50) + p(8000)$
		$p(8000) + \bar{p}\ (8000)$

SLAC (B FACTORY, BABAR)

Stanford Linear Accelerator Centre, located just south of San Francisco, is the longest linear accelerator in the world. It accelerates electrons and positrons down its 2-mile length to various targets, rings and detectors at its end. The PEP ring shown is being rebuilt for the B factory, which will study some of the mysteries of antimatter using B mesons. Related physics will be done at Cornell with CESR and in Japan with KEK.

FERMILAB (D0 AND CDF)

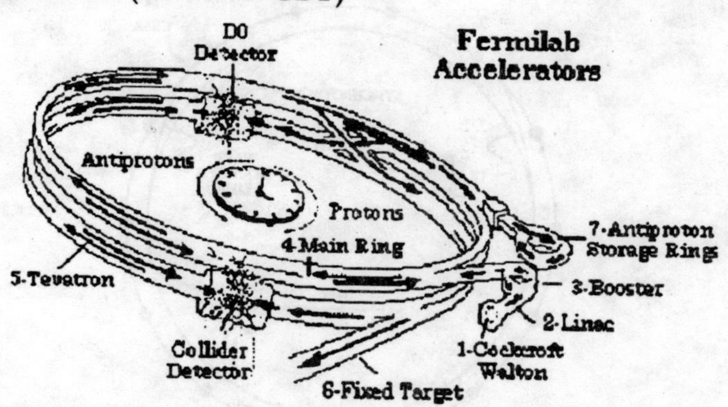

Fig. A picture of fermilab

Fermi National Accelerator Laboratory, a high-energy physics laboratory, named after particle physicist pioneer Enrico Fermi, is located 30 miles west of Chicago. It is the home of the world's most powerful particle accelerator, the Tevatron, which was used to discover the top quark.

Cern (LEP and LHC)

CERN (European Laboratory for Particle Physics) is an international laboratory where the W and Z bosons were discovered. CERN is the birthplace of the World-Wide Web. The Large Hadron Collider (see below) will search for Higgs bosons and other new fundamental particles and forces.

BROOKHAVEN (RHIC)

Brookhaven National Laboratory (BNL) is located on Long Island, New York. Charm quark was discovered there, simultaneously with SLAC. The main ring (RHIC) is 0.6 km in radius.

CORNELL (CESR)

The Cornell Electron-Positron Storage Ring (CESR) is an electron-positron collider with a circumference of 768 meters, located 12 meters below the ground at Cornell University campus. It is capable of producing collisions between electrons and their anti-particles, positrons, with centre-of-mass energies between 9 and 12 GeV.

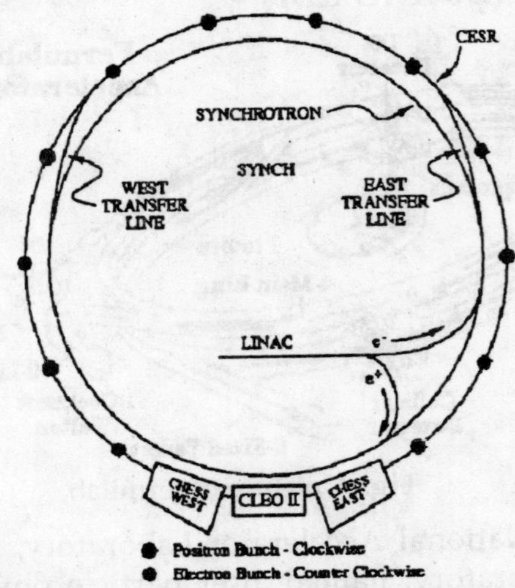

Figure: A picture of the Cornell accelerator

The products of these collisions are studied with a detection apparatus, called the CLEO detector.

DESY (HERA AND PETRA)

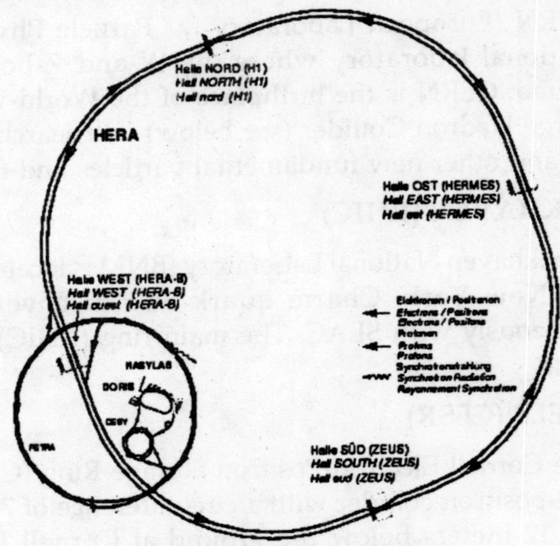

Fig. A picture of HERA

The DESY laboratory, located in Hamburg, Germany, discovered the gluon at the PETRA accelerator. DESY consists of two accelerators: HERA and PETRA. These accelerators collide electrons and protons.

KEK (TRISTAN)

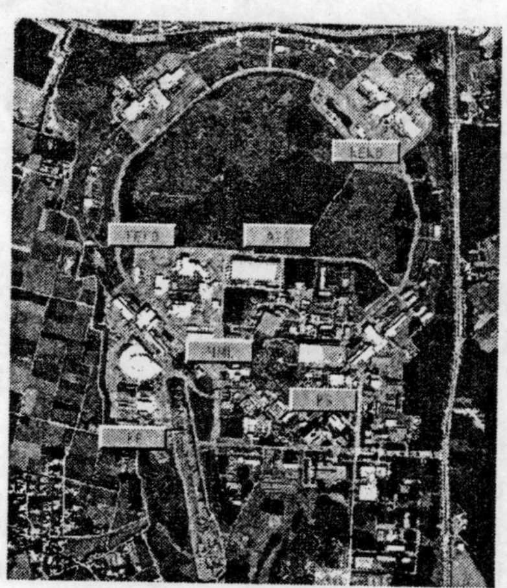

Fig. A picture of KEK

The KEK laboratory, in Japan, was originally established for the purpose of promoting experimental studies on elementary particles. A 12 GeV proton synchrotron was constructed as the first major facility. Since its commissioning in 1976, the proton synchrotron played an important role in boosting experimental activities in Japan and thus laid the foundation of the next step of KEK's high energy physics programme, a 30 GeV electron-positron colliding-beam accelerator called TRISTAN.

Chapter 22

Nuclear Masses

1. Each nucleus has a (positive) charge *Ze*, and integer number times the elementary charge *e*. This follows from the fact that atoms are neutral!
2. Nuclei of identical charge come in different masses, all approximate multiples of the "nucleon mass". (Nucleon is the generic term for a neutron or proton, which have almost the same mass, $m_p = 938.272\text{MeV}/c^2$, $m_n = \text{MeV}/c^2$.) Masses can easily be determined by analysing nuclei in a *mass spectrograph* which can be used to determine the relation between the charge Z (the number of protons, we believe) vs. the mass.

Nuclei of identical charge (chemical type) but different mass are called isotopes. Nuclei of approximately the same mass, but different chemical type, are called isobars.

MASS SPECTROGRAPH

A mass spectrograph is a combination of a bending magnet, and an electrostatic device (to be completed).

INTERPRETATION

We conclude that the nucleus of mass $m \approx Am_N$ contains Z positively charged nucleons (protons) and $N = A - Z$ neutral nucleons (neutrons). These particles are bound together by the "nuclear force", which changes the mass below that of free particles. We shall typically write ^{A}El for an element of chemical type ^{A}El, which determines *Z*, containing *A* nucleons.

DEEPER ANALYSIS OF NUCLEAR MASSES

To analyse the masses even better we use the atomic mass unit (amu), which is 1/12th of the mass of the neutral carbon atom,

$$1 \text{ amu} = \frac{1}{12} m_{^{12}\text{C}}.$$

This can easily be converted to SI units by some chemistry. One mole of ^{12}C weighs 0.012 kg, and contains Avogadro's number particles, thus

$$1 \text{ amu} = \frac{0.001}{N_A} \text{kg} = 1.66054 \times 10^{-27} \text{ kg} = 931.494 \text{MeV}/c^2.$$

The quantity of most interest in understanding the mass is the binding energy, defined for a neutral atom as the difference between the mass of a nucleus a ıd the mass of its constituents,

$$B(A, Z) = ZM_H c^2 + (A - Z)M_n c^2 - M(A, Z)c^2.$$

With this choice a system is bound when $B > 0$, when the mass of the nucleus is lower than the mass of its constituents. Let us first look at this quantity per nucleon as a function of A.

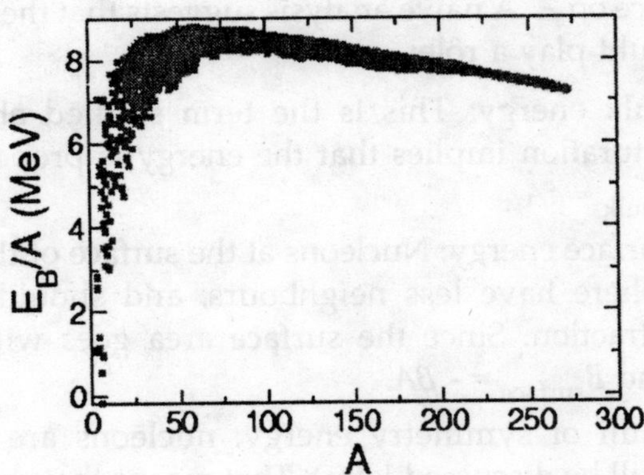

Fig. *B*/*A* versus *A*

This seems to show that to a reasonable degree of approximation the mass is a function of *A* alone, and

furthermore, that it approaches a constant. This is called nuclear saturation. This agrees with experiment, which suggests that the radius of a nucleus scales with the 1/3rd power of A,

$$R_{RMS} \approx 1.1 A^{1/3} \text{ fm}.$$

This is consistent with the saturation hypothesis made by Gamov in the 30's:

As A increase the volume per nucleon remains constant.

For a spherical nucleus of radius R we get the condition

$$\frac{4}{3}\pi R^3 = AV_1,$$

or

$$R = \left(\frac{V_1 3}{4\pi}\right)^{1/3} A^{1/3}.$$

From which we conclude that

$$V_1 = 5.5 \text{ fm}$$

NUCLEAR MASS FORMULA

There is more structure in Fig. than just a simple linear dependence on A. A naive analysis suggests that the following terms should play a rôle:

1. Bulk energy: This is the term studied above, and saturation implies that the energy is proportional to $B_{bulk} = aA$.
2. Surface energy: Nucleons at the surface of the nuclear sphere have less neighbours, and should feel less attraction. Since the surface area goes with R^2, we find $B_{surface} = -\beta A$.
3. Pauli or symmetry energy: nucleons are fermions (will be discussed later). That means that they cannot occupy the same states, thus reducing the binding. This is found to be proportional to $B_{symm} = -\gamma(N/2 - Z/2)^2/A^2$.

4. Coulomb energy: protons are charges and they repel. The average distance between is related to the radius of the nucleus, the number of interaction is roughly Z^2 (or $Z(Z - 1)$). We have to include the term $B_{Coul} = -\in Z^2/A$.

Taking all this together we fit the formula

$B(A, Z) = \alpha A - {}^{\beta}A^{2/3} - {}^{\gamma}(A/2 - Z)^2 A^{-2} - {}^{\varepsilon}Z^2 A^{-1/3}$

to all know nuclear binding energies with $A \geq 16$ (the formula is not so good for light nuclei). The fit results are given in table.

Table: Fit of masses to Eq.

Parameter	Value
α	15.36 MeV
β	16.32 MeV
γ	90.45 MeV
ε	0.6928 MeV

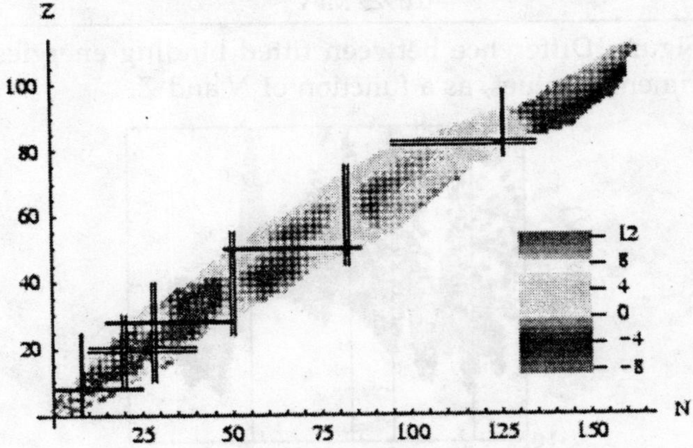

Fig. Difference between fitted binding energies and experimental values, as a function of *N* and Z.

In Fig. we show how well this fit works. There remains a certain amount of structure, see below, as well as a strong difference between neighbouring nuclei. This is due to the superfluid nature of nuclear material: nucleons of opposite momenta tend to anti-align their spins, thus gaining energy. The solution is to add a pairing term to the binding energy,

$$B_{\text{pair}} = \begin{cases} \delta A^{-1/2} & \text{for } N \text{ even, } Z \text{ odd} \\ -\delta A^{-1/2} & \text{for } N \text{ odd, } Z \text{ even} \end{cases}$$

The results including this term are significantly better, even though all other parameters remain at the same position, see Table Taking all this together we fit the formula

$$B(A, Z) = \alpha A - \beta A^{2/3} - \gamma(A/2 - Z)^2 A^{-2} - \delta B_{\text{pair}}(A, Z) - \varepsilon Z^2 A^{-1/3}$$

Table: Fit of masses to Eq.

Parameter	Value
α	15.36 MeV
β	16.32 MeV
γ	90.46 MeV
δ	11.32 MeV
ε	0.6929 MeV

Figure: Difference between fitted binding energies and experimental values, as a function of N and Z.

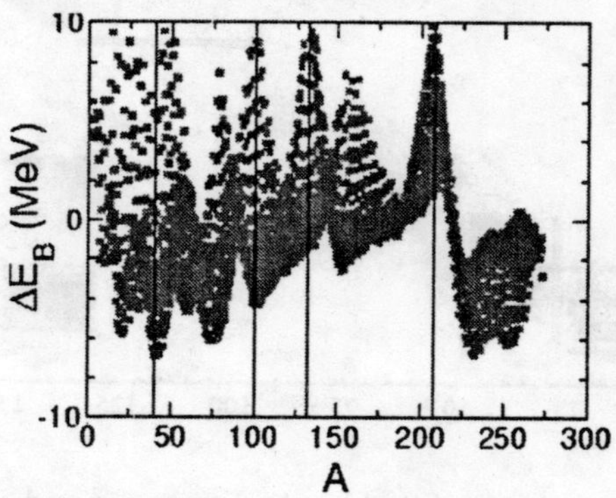

Fig. B/A **versus** A**, mass formula subtracted**

STABILITY OF NUCLEI

In figure we have colour coded the nuclei of a given mass $A = N + Z$ by their mass, red for those of lowest mass through to magenta for those of highest mass. We can see that typically the nuclei that are most stable for fixed *A* have more neutrons than protons, more so for large *A* increases than for low *A*. This is the "neutron excess".

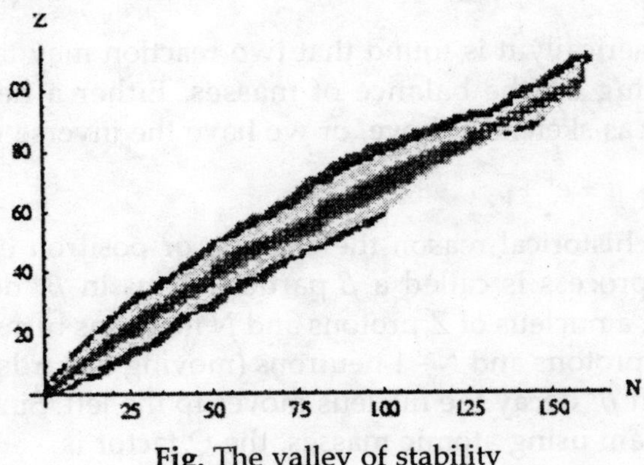

Fig. The valley of stability

β DECAY

If we look at fixed nucleon number *A*, we can see that the masses vary strongly,

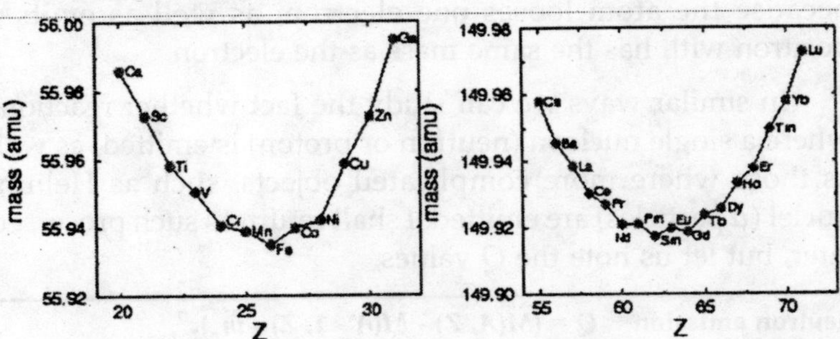

Fig. A cross section through the mass table for fixed *A*. To the left, $A = 56$, and to the right, $A = 150$.

It is known that a free neutron is not a stable particle, it actually decays by emission of an electron and an antineutrino,

$$n \rightarrow p + e^- + \bar{\nu}_e .$$

The reason that this reaction can take place is that it is endothermic, $m_n c^2 > m_p c^2 + m_e c^2$. (Here we assume that the neutrino has no mass.) The degree of allowance of such a reaction is usually expressed in a Q value, the amount of energy released in such a reaction,

$$Q = m_{\mathrm{n}}c^2 - m_{\mathrm{p}}c^2 - m_{\mathrm{e}}c^2 = 939.6 - 938.3 - 0.5 = 0.8 \text{ MeV}.$$

Generically it is found that two reaction may take place, depending on the balance of masses. Either a neutron "β decays" as sketched above, or we have the inverse reaction

$$p \rightarrow n + e^+ + \bar{\nu}_e.$$

For historical reason the electron or positron emitted in such a process is called a β particle. Thus in β^- decay of a nucleus, a nucleus of Z protons and N neutrons turns into one of $Z + 1$ protons and $N - 1$ neutrons (moving towards the right in Fig. In β^+ decay the nucleus moves to the left. Since in that figure I am using atomic masses, the Q factor is

$$Q\beta^- = \quad M(A, Z)c^2 - M(A, Z + 1)c^2,$$

$$Q\beta^- = \quad M(A, Z)c^2 - M(A, Z - 1)c^2 - 2m_e c^2.$$

The double electron mass contribution in this last equation because the atom looses one electron, as well as emits a positron with has the same mass as the electron.

In similar ways we can study the fact whether reactions where a single nucleon (neutron or proton) is emitted, as well as those where more complicated objects, such as Helium nuclei (α particles) are emitted. I shall return to such processed later, but let us note the Q values,

neutron emission	$Q = (M(A, Z) - M(A - 1, Z) - m_{\mathrm{n}})c^2,$
proton emission	$Q = (M(A, Z) - M(A - 1, Z - 1) - M(1, 1))c^2,$
emission	$Q = (M(A, Z) - M(A - 4, Z - 2) - M(4, 2))c^2,$
break up	$Q = (M(A, Z) - M(A - A_1, Z - Z_1) - M(A_1, Z_1))c^2.$

Nuclei are quantum systems, and as such must be described by a quantum Hamiltonian. Fortunately nuclear energies are much smaller than masses, so that a description in terms of non-relativistic quantum mechanics is possible. Such a description is not totally trivial since we have to deal with quantum systems containing many particles. Rather then solving such complicated systems, we often resort to models.

Chapter 23

Quantum Numbers

As in any quantum system there are many quantum states in each nucleus. These are labelled by their quantum numbers, which, as will be shown later, originate in symmetries of the underlying Hamiltonian, or rather the underlying physics.

ANGULAR MOMENTUM

One of the key invariances of the laws of physics is rotational invariance, i.e., physics is independent of the direction you are looking at. This leads to the introduction of a vector angular momentum operator,

$$\hat{L} = \hat{r} \times \hat{p},$$

which generates rotations. As we shall see later quantum states are not necessarily invariant under the rotation, but transform in a well-defined way. The three operators $\hat{L}x, \hat{L}y$, and $\hat{L}x$ satisfy a rather intriguing structure,

$$[\hat{L}x, \hat{L}y] = \hat{L}x\hat{L}y - \hat{L}y\hat{L}x = i\hbar\,\hat{L}x,$$

and the same for q cyclic permutation of indices ($xyz \rightarrow yzx$ or zxy). This shows that we cannot determine all three components simultaneously in a quantum state. One normally only calculates the length of the angular momentum vector, and its projection on the z axis,

$$\hat{L}^2 \phi LM = \hbar^2 L(L+1)$$

$$\hat{L}_z\phi LM = \hbar Lz\phi LM$$

It can be shown that L is a non-negative integer, and M is an integer satisfying $|M|<L$, i.e., the projection is always smaller than or equal to the length, a rather simple statement in classical mechanics.

The standard, albeit slightly simplified, picture of this process is that of a fixed length angular momentum precessing about the z axis, keeping the projection fixed, as shown in Fig.

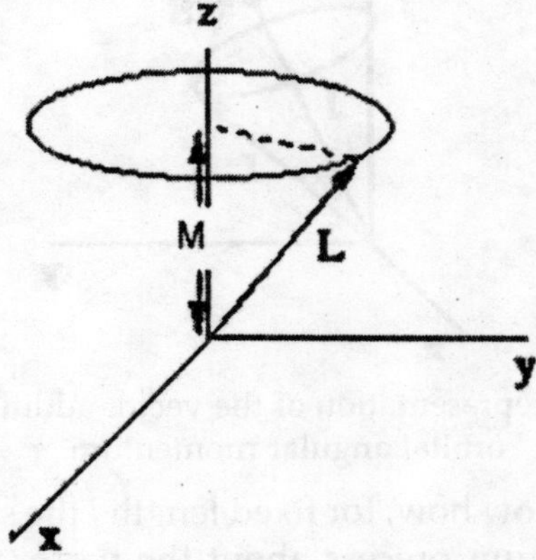

Fig. A pictorial representation of the "quantum precession" of an angular momentum of fixed length L and projection M.

The energy of a quantum state is independent of the M quantum number, since the physics is independent of the orientation of L in space (unless we apply a magnetic field that breaks this symmetry). We just find multiplets of $2L + 1$ states with the same energy and value of L, differing only in M.

Unfortunately the story does not end here. Like electrons, protons and neutron have a spin, i.e., we can use a magnetic field to separate nucleons with spin up from those with spin down. Spins are like orbital angular momenta in many aspects, we can write three operators $\hat{S}$ that satisfy the same relation as the $\hat{L}$'s, but we find that

$$\hat{S}^2\phi S,SSz = \hbar^2\frac{3}{4}\phi S,Sz$$

i.e., the length of the spin is 1/2, with projections ±1/2.

Spins will be shown to be coupled to orbital angular momentum to total angular momentum J,

$$\hat{J} = \hat{L} + \hat{S},$$

and we shall specify the quantum state by L, S, J and J_z. This can be explained pictorially as in Fig. .

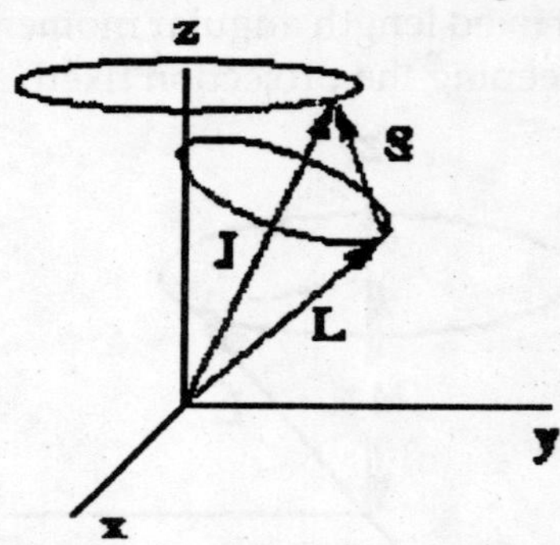

Fig. A pictorial representation of the vector addition of spin and orbital angular momentum.

There we show how, for fixed length J the spin and orbital angular momentum precess about the vector J, which in its turn precesses about the z-axis. It is easy to see that if *vecL* and S are fully aligned we have $J = L + S$, and if they are anti-aligned $J = | L - S |$. A deeper quantum analysis shows that this is the way the quantum number work. If the angular momentum quantum numbers of the states being coupled are L and S, the length of the resultant vector J can be

$$J = | L - S |, | L - S | + 2, \ldots, L + S.$$

We have now discussed the angular momentum quantum number for a single particle. For a nucleus which in principle is made up from many particles, we have to add all these angular momenta together until we get something called the total angular momentum. Since the total angular momentum of a single particle is half-integral (why?), the total angular momentum of a nucleus is integer for even A, and half-integer for odd A.

PARITY

Another symmetry of the wave function is parity. If we change $r \rightarrow r$, i.e., mirror space, the laws of physics are invariant. Since we can do this operation twice and get back where we started from, any eigenvalue of this operation must be ±1, usually denoted as $\Pi = \pm$. It can be shown that for a particle with orbital angular momentum L, $\Pi = (-1)^L$. The parity of many particles is just the product of the individual parities.

ISOTOPIC SPIN (ISOBARIC SPIN, ISOSPIN)

The most complicated symmetry in nuclear physics is isospin. In contrast to the symmetries above this is not exact, but only approximate. The first clue of this symmetry come from the proton and neutron masses, m_n = MeV/c^2 and m_p = MeV/c^2, and their very similar behaviour in nuclei. Remember that the dominant binding terms only depended on the number of nucleons, not on what type of nucleons we are dealing with.

All of this leads to the assumption of another abstract quantity, called isospin, which describes a new symmetry of nature. We assume that both neutrons and protons are manifestation of one single particle, the nucleon, with isospin down or up, respectively. We shall have to see whether this makes sense by looking in more detail at the nuclear physics. We propose the identification

$$Q = (I_z + 1/2)e,$$

where I_z is the z projection of the vectorial quantity called isospin. Apart from the neutron-proton mass difference, isospin symmetry in nuclei is definitely broken by the Coulomb force, which acts on protons but not on neutrons. We shall argue that the nuclear force, that couples to the "nucleon charge" rather than electric charge, respects this symmetry. What we shall do is look at a few nuclei where we can study both a nucleus and its mirror image under the exchange of protons and neutrons. One example are the nuclei ^{7}He and ^{7}B (2 protons and 5 neutrons, $I_z = -3/2$ vs. 5 protons and 2 neutrons, $I_z = 3/2$) and ^{7}Li and ^{7}B (3 protons and 4 neutrons, $I_z = -1/2$ vs. 4 protons and 3 neutrons, $I_z = 1/2$).

We note there the great similarity between the pairs of mirror nuclei. Of even more importance is the fact that the 3/2$^-$;3/2 level occurs at the same energy in all four nuclei, suggestion that we can define these states as an "isospin multiplet", the same state just differing by I_z

DEUTERON

Let us think of the deuteron (initially) as a state with $L = 0$, $J = 1$, $S = 1$, usually denoted as $^3S_1$ (S means $L = 0$, the 3 denotes $S = 1$, i.e., three possible spin orientations, and the subscript 1 the value of J). Let us model the nuclear force as a three dimensional square well with radius R. The Schrodinger equation for the spherically symmetric S state is (work in radial coordinates)

$$-\frac{\hbar^2}{2\mu}\frac{1}{r^2}\left(\frac{d}{dr}r^2\frac{d}{dr}\right)R(r) + V(r)R(r) = ER(r).$$

Here $V(r)$ is the potential, and μ is the reduced mass,

$$\mu = \frac{m_n m_p}{m_n + m_p},$$

which arrises from working in the relative coordinate only. It is easier to work with $u(r) = rR(r)$, which satisfies the condition

$$\frac{\hbar}{2\mu}\frac{d^2}{dr^2} - u(r) + V(r)u(r) = Eu(r),$$

as well as $u(0) = 0$. The equation in the interior

$$-\frac{\hbar}{2\mu}\frac{d^2}{dr^2}\ u(r)V_0u(r) = Eu(r), \qquad u(0) = 0$$

has as solution

$$u = A\ \sin kr, \qquad k = \sqrt{\frac{2\mu}{\hbar^2}(V_0 + E)}.$$

Outside the well we find the standard damped exponential,

$$u = B\exp(kr), \qquad k = \sqrt{\frac{2\mu}{\hbar^2}(-E)}\,.$$

Matching derivatives at the boundary we find

$$-\cot kR = \frac{K}{k} = \sqrt{\frac{-E}{V_0 + E}}\,.$$

We shall now make the assumption that $|E| \ll V_0$, which will prove true. Then we find

$$k \approx \sqrt{\frac{2\mu}{\hbar}V_0}\,, \qquad \cot k\,R \approx 0.$$

Since it is known from experiment that the deuteron has only one bound state at energy -2.224573±0.000002 MeV, we see that $kR \approx \pi/2$! Substituting k we see that

$$V_0 R^2 = \frac{\pi^2\hbar^2}{8\mu}.$$

If we take V_0 = 30 MeV, we find R = 1.83 fm.

We can orient the spins of neutron and protons in a magnetic field, i.e., we find that there is an energy

$E_{magn} = \mu\,N\,\mu\,S\cdot B.$

(The units for this expression is the so-called nuclear magneton, $\mu N = \dfrac{e\hbar}{2m_p}$.) Experimentally we know that

$$\mu_n = -\,1.91315 \pm 0.00007\mu_N \qquad \mu_p = 2.79271 \pm 0.00002\,\mu_N.$$

If we compare the measured value for the deuteron, μ_d= 0.857411±0.000019μ_N, with the sum of protons and neutrons (spins aligned), we see that $\mu_p + \mu_n$= 0.857956±0.00007μ_N. The close agreement suggest that the spin assignment is largely OK; the small difference means that our answer cannot be the whole story: we need other components in the wave function.

We know that an S state is spherically symmetric and cannot have a quadrupole moment, i.e., it does not have a preferred axis of orientation in an electric field. It is known that the deuteron has a positive quadrupole moment of $0.29e^2$ fm^2, corresponding to an elongation of the charge distribution along the spin axis.

From this we conclude that the deuteron wave function carries a small (7 per cent) component of the $^3D_1$ state (D: L = 2). We shall discuss later on what this means for the nuclear force.

SCATTERING OF NUCLEONS

We shall concentrate on scattering in an $L = 0$ state only, further formalism just gets too complicated. For definiteness I shall just look at the scattering in the $^3S_1$ channel, and the $^1S_0$ one. (These are also called the triplet and singlet channels.)

NUCLEAR FORCES

Having learnt this much about nuclei, what can we say about the nuclear force, the attraction that holds nuclei together? First of all, from Rutherford's old experiments on particle scattering from nuclei, one can learn that the range of these forces is a few fm.

From the fact that nuclei saturate, and are bound, we would then naively build up a picture of a potential that is strongly repulsive at short distances, and shows some mild attraction at a range of 1-2 fm, somewhat like sketched in Fig.

Here we assume, that just as the Coulomb force can be derived from a potential that only depends on the size of r,

$$V(r) = \frac{q_1 q_2}{4\pi \epsilon_0 r},$$

the nuclear force depends only on r as well. This is the simplest way to construct a rotationally invariant energy. For particles with spin other possibilities arise as well (e.g., $\hat{S} \cdot \mathrm{r}$) so how can we see what the nuclear force is really like?

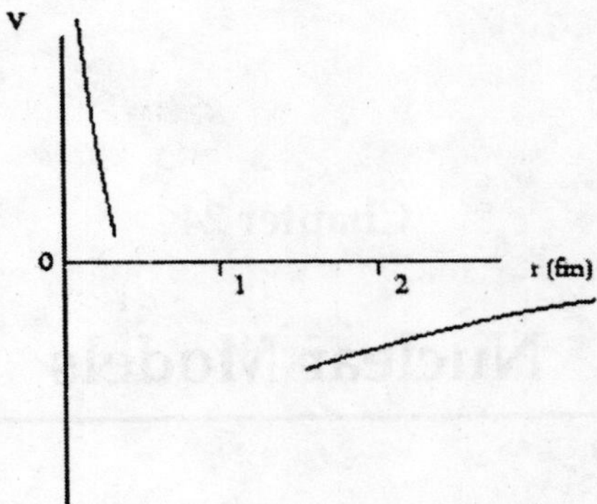

Fig. Possible form for the internucleon potential, repulsive at short distances, and attractive at large distances.

Since we have taken the force to connect pairs of particles, we can just study the interaction of two nucleons, by looking both at the bound states (there is only one), and at scattering, where we study how a nucleon gets deflected when it scatters of another nucleon. Let us first look at the deuteron, the bound state of a proton and a neutron. The quantum numbers of its ground state are $J\pi = 1^+$, $I = 0$. A little bit of additional analysis shows that this is a state with $S = 1$, and $L = 0$ or 2. Naively one would expect a lowest state $S = 0$, $L = 0$ (which must have $I =$ for symmetry reasons not discussed here). So what can we read of about the nuclear force from this result?

We conclude the following:

1. The nuclear force in the *S* waves is attractive.
2. Nuclear binding is caused by the tensor force.
3. The nuclear force is isospin symmetric (i.e., it is independent of the direction of isospin).

Chapter 24

Nuclear Models

There are two important classed of nuclear models: single particle and microscopic models, that concentrate on the individual nucleons and their interactions, and collective models, where we just model the nucleus as a collective of nucleons, often a nuclear fluid drop.

Microscopic models need to take into account the Pauli principle, which states that no two nucleons can occupy the same quantum state. This is due to the Fermi-Dirac statistics of spin 1/2 particles, which states that the wave function is antisymmetric under interchange of any two particles.

NUCLEAR SHELL MODEL

The simplest of the single particle models is the nuclear shell model. It is based on the observation that the nuclear mass formula, which describes the nuclear masses quite well on average, fails for certain "magic numbers", i.e., for neutron number $N = 20, 28, 50, 82, 126$ and proton number $Z = 20, 28, 50, 82$, as indicated if Fig. xxxx. These nuclei are much more strongly bound than the mass formula predicts, especially for the doubly magic cases, i.e., when N and Z are both magic. Further analysis suggests that this is due to a shell structure, as has been seen in atomic physics.

MECHANISM THAT CAUSES SHELL STRUCTURE

So what causes the shell structure? In atoms it is the Coulomb force of the heavy nucleus that forces the electrons to occupy certain orbitals. This can be seen as an external agent.

In nuclei no such external force exits, so we have to find a different mechanism.

The solution, and the reason the idea of shell structure in nuclei is such a counter-intuitive notion, is both elegant and simple. Consider a single nucleon in a nucleus. Within this nuclear fluid we can consider the interactions of each of the nucleons with the one we have singled out. All of these nucleons move rather quickly through this fluid, leading to the fact that our nucleons only sees the average effects of the attraction of all the other ones. This leads to us replacing, to first approximation, this effect by an average nuclear potential, as sketched in Fig..

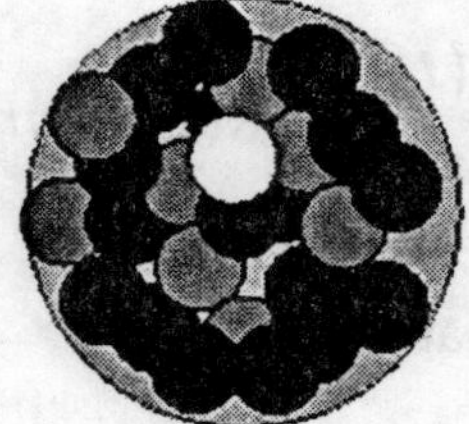

Fig. A sketch of the averaging approximation

Thus the idea is that the shell structure is caused by the average field of all the other nucleons, a very elegant but rather surprising notion!

MODELING THE SHELL STRUCTURE

Whereas in atomic physics we solve the Coulomb force problem to get the shell structure, we expect that in nuclei the potential is more attractive in the centre, where the density is highest, and less attractive near the surface. There is no reason why the attraction should diverge anywhere, and we expect the potential to be finite everywhere. One potential that satisfies these criteria, and can be solved analytically, is the Harmonic oscillator potential. Let us use that as a first model, and solve

$$\frac{\hbar}{2m}\Delta\psi(r)+\frac{1}{2}m\omega^2 r^2(r)=E\psi(r).$$

The easiest way to solve this equation is to realise that, since $r^2 = x^2 + y^2 + z^2$, the Hamiltonian is actually a sum of an x, y and z harmonic oscillator, and the eigenvalues are the sum of those three oscillators,

$$E_{nxnynz} = (n_x + n_y + n_z + 3/2)\hbar\omega .$$

The great disadvantage of this form is that it ignores the rotational invariance of the potential. If we separate the Schrodinger equation in radial coordinates as

$$R_{nl}(r)Y_{LM}(\theta,\phi)$$

with Y the spherical harmonics, we find

$$-\frac{\hbar^2}{2m}\frac{1}{r^2}\frac{\partial}{\partial r}\left(r^2\frac{\partial}{\partial r}R(r)\right) + \frac{\hbar^2 L(L+1)}{r^2}R(r) + mr^2R(r) = E_{nL}R(r).$$

In this case it can be shown that

$$E_{nL} = (2n + L + 3/2)\hbar\omega ,$$

and L is the orbital angular momentum of the state. We use the standard, so-called spectroscopic, notation of $s, p, d, f, g, h, i, j,...$ for $L = 0, 1, 2,....$

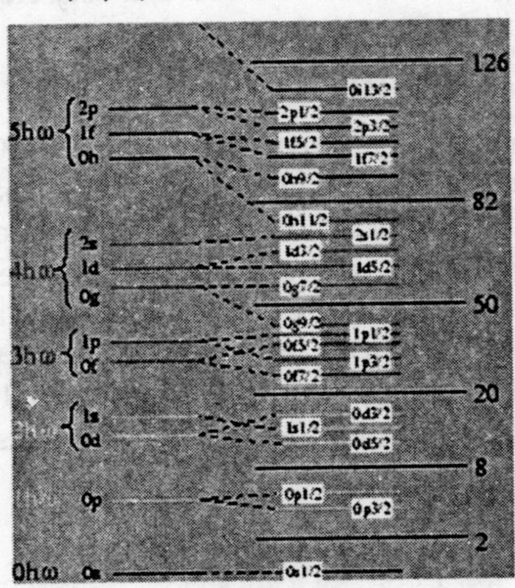

Fig. A schematic representation of the shell structure in nuclei.

In the left hand side of Fig. we have list the number harmonic oscillator quanta in each set of shells. We have made use of the fact that in the real potentials state of the same number of quanta but different *L* are no longer degenerate, but there are groups of shells with big energy gaps between them. This cannot predict the magic numbers beyond 20, and we need to find a different mechanism. There is one already known for atoms, which is the so-called spin orbit splitting. This means that the degeneracy in the total angular momentum ($j = L \pm 1/2$) is lifted by an energy term that splits the aligned from anti-aligned case. This is shown schematically in the right of the figure, where we label the states by *n*, *l* and *j*. The gaps between the groups of shells are in reality much larger than the spacing within one shell, making the binding-energy of a closed-shell nucleus much lower than that of its neighbours.

EVIDENCE FOR SHELL STRUCTURE

Evidence for the shell structure can be seen in two ways:

1- By looking at nuclear reactions that add a nucleon or remove a nucleon from a closed shell nucleus. The most sensitive of these are electron knockout reactions, where an electron comes in and an electron and a proton or neutron escapes, usually denoted as (*e*, *e′p*) (*e*, *e′n*) reactions. In those we see clear evidence of peaks at the single particle energies.

2- By looking at nuclei one particle or one hole away from a doubly magic nucleus. As an example look at the nuclei around 208Pb, as in Fig.

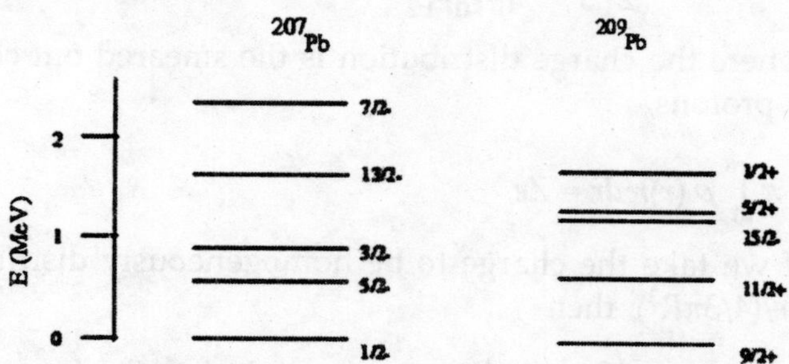

Fig. The spectra for the one-neutron hole nucleus ^{207}Pb and the one-particle nucleus ^{209}Pb.

In order to understand this figure we need to think a little about the shell structure, as sketched very schematically in Fig.: The one neutron-hole nucleus corresponds to taking away a single neutron from the 50-82 shell, and the one neutron particle state to adding a a neutron above the $N = 128$ shell closure. We can also understand more clearly why a closed shell nucleus has very few low-energy excited quantum states, since we would have to create a hole below the closed shell, and promote the nucleon in that shell to an open state above the closure. This requires an energy that equals the gap in the single-particle energy.

COLLECTIVE MODELS

Another, and actually older, way to look at nuclei is as a drop of "quantum fluid". This ignores the fact that a nucleus is made up of protons and neutrons, and explains the structure of nuclei in terms of a continuous system, just as we normally ignore the individual particles that make up a fluid.

LIQUID DROP MODEL AND MASS FORMULA

Now we have some basic information about the liquid drop model, let us try to reinterpret the mass formula in terms of this model; especially as those of a spherical drop of liquid.

As a prime example consider the Coulomb energy. The general energy associated with a charge distribution is

$$E_{\text{Coulomb}} = \frac{1}{2} \int \frac{\rho(r_1)\rho(r_2)}{4\pi\epsilon_0 r_{12}} d^3r_1 d^3r_{2'}$$

where the charge distribution is the smeared out charge of the protons,

$$4\pi \int_0^R \rho(r) r^2 dr = Ze$$

If we take the charge to be homogeneously distributed $\rho = Ze/(4/3\pi R^3)$, then

$$E_{\text{Coulomb}} = \frac{(2\pi)(4\pi)}{4\pi \epsilon_0} p^2 \int_0^R \int_{-1}^{1} \frac{r_1^2 dr_2 d\cos\theta r_2^2 dr_2}{(r_1^2 + r_2^2 - 2r_1 r_2 \cos\theta)^{1/2}}$$

$$= \frac{\pi\rho^2}{2\epsilon_0}\int_0^R\int_0^R \frac{(r_1+r_2)-|r_1+r_2|}{r_1r_2} r_1{}^2dr_1\, r_2{}^2dr_2$$

$$= \frac{\pi\rho^2}{2\epsilon_0}[2\int_0^R x^2dx\int_0^R ydy - \int_0^R\int_0^R |r_1-r_2|r_1r_2dr_1dr_2]$$

$$= \frac{\pi\rho^2}{2\epsilon_0}\frac{R^5}{3} - R^5 15$$

$$= \frac{\pi Z^2 9e^2}{(4\pi)^2 R^6 2\epsilon_0}\frac{4R^5}{15} = \frac{e^2}{4\pi\epsilon_0 \frac{3}{10}}\frac{Z^2}{R}$$

EQUILIBRIUM SHAPE AND DEFORMATION

Once we picture a nucleus as a fluid, we can ask question about its equilibrium shape. From experimental data we know that near closed shells nuclei are spherical, i.e., the equilibrium shape is a sphere. When both the proton and neutron number differ appreciably from the magic numbers, the ground state is often found to be axially deformed, either prolate (cigar like) or oblate (like a pancake).

A useful analysis to perform is to see what happens when we deform a nucleus slightly, turning it into an ellipsoid, with one axis slightly longer than the others, keeping a constant volume:

$a = R(1 + \varepsilon)$, $\quad b = (R(1 + \varepsilon)^{-1/2}$.

The volume is $\frac{4}{3}\pi\, ab^2$, and is indeed constant. The surface area of an ellipsoid is more complicated, and we find

$$S = 2\pi[b^2 + ab, \frac{\arcsin e}{e}]$$

where the eccentricity e is defined as

$e = [(1 - b^2/a^2]^{1/2}$.

For small deformation we find a much simpler result,

$$S = 4\pi R^2[1 + \frac{2}{5}\in^2],$$

and the surface area thus increases for both elongations and contractions. Thus the surface energy increases by the same factor. There is one competing term, however, since the Coulomb energy also changes, the Coulomb energy goes down, since the particles are further apart,

$$E_{\text{Coulomb}} \rightarrow E_{\text{Coulomb}}(1 - \frac{\in^2}{5})$$

We thus find a change in energy of

$$DeltaE = \in^2 [\frac{2}{5}\beta A^{2/3} - \frac{1}{5}\in Z^2 A^{-1/3}$$

The spherical shape is stable if $\Delta E > 0$.

Since it is found that the nuclear fluid is to very good approximation incompressible, the dynamical excitations are those where the shape of the nucleus fluctuates, keeping the volume constant, as well as those where the nucleus rotates without changing its intrinsic shape.

COLLECTIVE VIBRATIONS

Let us first look at collective vibrations, and for simplicity only at those of a spherical fluid drop. We can think of a large number of shapes; a complete set can be found by parametrising the surface as $r = \sum L, M\, a_{LM} Y_{LM}(\theta, \phi)$, where Y_{LM} are the spherical harmonics and describe the multipolarity (angular momentum) of the surface. A few examples are shown in Figs. xxx, where we sketch the effects of monopole ($L = 0$), dipole ($L = 1$), quadrupole ($L = 2$) and octupole ($L = 3$) modes. Let us investigate these modes in turn, in the harmonic limit, where we look at small vibrations (small a_{LM}) only.

MONOPOLE

The monopole mode, see Fig., is the one where the size of the nuclear fluid oscillates, i.e., where the nucleus gets compressed. Experimentally one finds that the lowest excitation of this type, which in even-even nuclei carries the quantum number $J\pi = 0^+$, occurs at an energy of roughly.

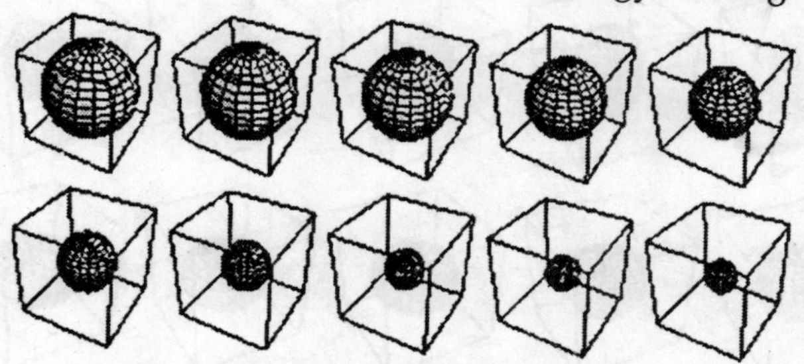

Fig. Monopole fluctuations of a liquid drop

$$E_0 \approx 80A^{-1/3}\text{MeV}$$

above the ground state. Compared to ordinary nuclear modes, which have energies of a few MeV, these are indeed high energy modes (15 MeV for A = 216), showing the incompressibility of the nuclear fluid.

DIPOLE

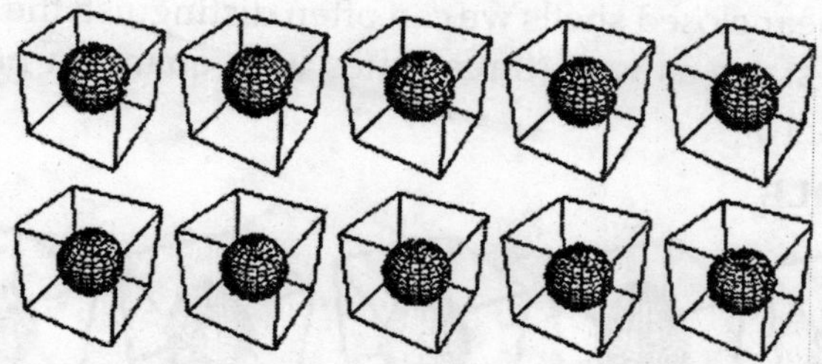

Fig. Dipole fluctuations of a liquid drop

The dipole mode, Fig. by itself is not very interesting: it corresponds to an overall translation of the centre of the nuclear fluid. One can, however, imagine a two-fluid model where a proton and neutron fluid oscillate against each other. This is a collective isovector (I = 1) mode. It has quantum numbers $J\pi = 1^-$, occurs at an energy of roughly

$$E_0 \approx 77A^{-1/3}\text{MeV}$$

above the ground state, close to the monopole resonance. It shows that the neutron and proton fluids stick together quite strongly, and are hard to separate.

QUADRUPOLE

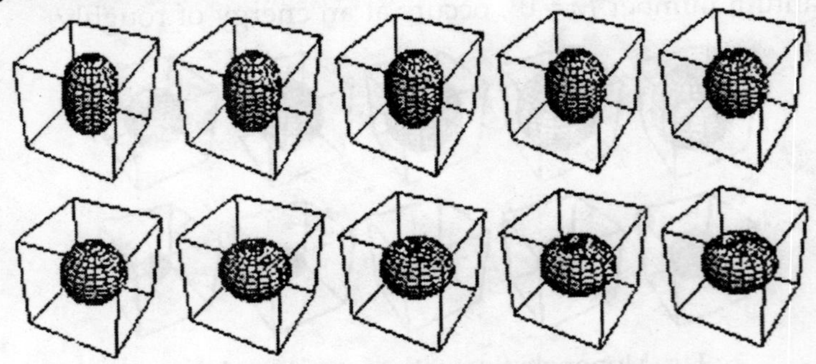

Fig. Quadrupole fluctuations of a liquid drop

Quadrupole modes, see Fig. , are the dominant vibrational feature in almost all nuclei. The very special properties of the lower multipolarities mean that these are the first modes available for low-energy excitations in nuclei. In almost all even-even nuclei we find a low-lying state (at excitation energy of less than 1 - 2MeV), which carries the quantum numbers $J\pi = 2^+$, and near closed shells we can often distinguish the second harmonic states as well (three states with quantum numbers $J\pi = 0^+, 2^+, 4^+$).

OCTUPOLE

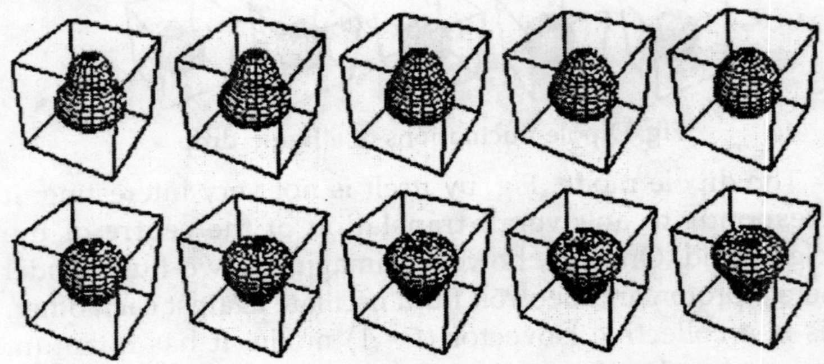

Fig. Octupole fluctuations of a liquid drop

Octupole modes, with $J = 3^-$, see Fig. can be seen in many nuclei. In nuclei where shell-structure makes quadrupole modes occur at very high energies, such as doubly magic nuclei, the octupole state is often the lowest excited state.

COLLECTIVE ROTATIONS

Fig. Collective rotation of an axially deformed liquid drop

Once we have created a nucleus with axial deformation, i.e., a nucleus with ellipsoidal shape, but still axial symmetry about one axis, we can rotate the fluid around one of the non-symmetry axes to generate excitations, see Fig. We cannot do it around a symmtery axis, since the resulting state would just be the same quantum state as we started with, and therefore the energy cannot change. A rotated state around a non-symmetry axis is a different quantum state, and therefore we can overlay many of these states, especially with constant rotational velocity. This is almost like the rotation of a dumbbell, and we can predict the classical spectrum to be of the form

$$H = \frac{1}{2L} \mathrm{J}^2,$$

where J is the classical angular momentum. We predict a quantum mechanical spectrum of the form

$$E_{\mathrm{rot}}(J) = \frac{\hbar^2}{2L} J(J+1),$$

where J is now the angular momentum quantum number. Naively we expect the spectrum to be more compressed (the moment of inertial is larger) the more elongated the nucleus becomes. It is known that certain structures in nuclei indeed describe well deformed nuclei, up to super and hyper deformed (axis ratio from 1: 1.2 to 1: 2).

FISSION

Once we have started to look at the liquid drop model, we can try to ask the question what it predicts for fission, where one can use the liquid drop model to good effect. We are studying how a nuclear fluid drop separates into two smaller ones, either about the same size, or very different in size.

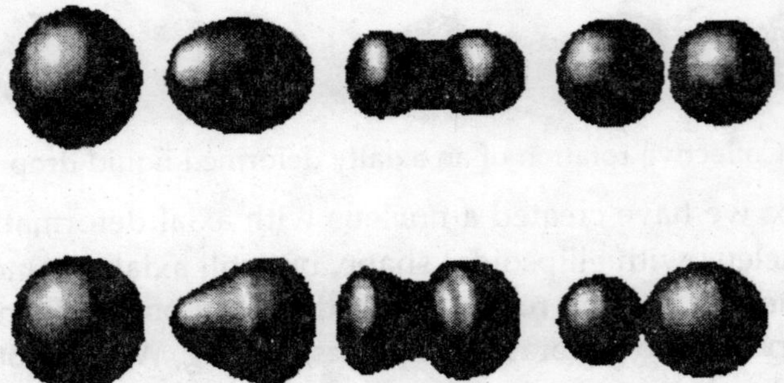

Fig. Symmetric (upper row) or asymmetric fission (lower row)

This process is indicated in Fig. The liquid drop elongates, by performing either a quadrupole or octupole type vibration, but it persists until the nucleus falls apart into two pieces. Since the equilibrium shape must be stable against small fluctuations, we find that the energy must go up near the spherical form, as sketched in Fig. .

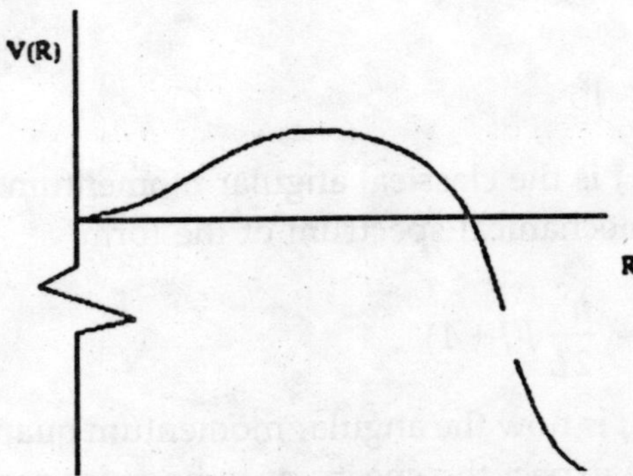

Fig. Potential energy for fission

In that figure we sketch the energy - which is really the potential energy - for separation into two fragments, R is the fragment distance. As with any of such processes we can either consider classical fission decays for energy above the fission barrier, or quantum mechanical tunneling for energies below the barrier. The method used in fission bombs is to use the former, by hitting a ^{235}U nucleus with a slow neutron a state with energy above the barrier is formed, which fissions fast. The fission products are unstable, and emit additional neutrons, which can give rise to a chain reaction.

The mass formula can be used to give an indication what is going on; Let us look at at the symmetric fusion of a nucleus. In that case the Q value is

$$Q = M(A, Z) - 2M(A/2, Z/2)$$

Please evaluate this for ^{236}U (92 protons). The mass formula fails in prediciting the asymmetry of fission, the splitting process is much more likely to go into two unequal fragments.

BARRIER PENETRATION

In order to understand quantum mechanical tunneling in fission it makes sense to look at the simplest fission process: the emission of a He nucleus, so called α radiation. The picture is as in Fig..

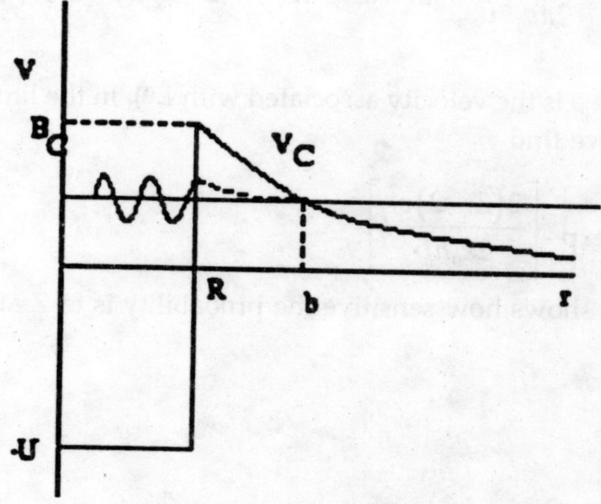

Fig. The potential energy for alpha decay

Suppose there exists an particle inside a nucleus at an (unbound) energy > 0. Since it isn't bound, why doesn't it decay immediately? This must be tunneling. In the sketch above we have once again shown the nuclear binding potential as a square well, but we have included the Coulomb tail,

$$V_{\text{Coulomb}}(r) = \frac{(Z-2)2e^2}{4\pi \in_0 r}.$$

The hight of the barrier is exactly the coulomb potential at the boundary, which is the nuclear readius, $R_C = 1.2A^{1.3}$ fm, and thus $B_C = 2.4(Z - 2)A^{-1/3}$. The decay probablility across a barrier can be given by the simple integral expression $P = e^{-2}$, with

$$\gamma = \frac{(2\mu\alpha)^{1/2}}{\hbar} \int_{RC}^{b} \left[V(r) - E^{\alpha}\right]^{1/2} dr$$

$$= \frac{(2\mu\alpha)^{1/2}}{\hbar} \int_{RC}^{b} \left[\frac{2(Z-2)}{4\pi\varepsilon_0 r} - E\alpha\right]^{1/2} dr$$

$$= \frac{2(Z-2)e^2}{2\pi\varepsilon_0 \hbar \upsilon} [\arccos(E^{\alpha}/B_C) - (E^{\alpha}/B_C)(1 - E^{\alpha}/B_C)],$$

(here v is the velocity associated with E^{α}). In the limit that $B_C \gg E$ we find

$$P = \exp - \left[\frac{2(Z-2)e^2}{2\varepsilon_0 \hbar \upsilon}\right].$$

This shows how sensitive the probability is to Z and v!

Chapter 25

The Fundamental Forces

The fundamental forces are normally divided in four groups, of the four so-called "fundamental" forces. These are often naturally classified with respect to a dimensionless measure of their strength. To set these dimensions we use $\hbar$, c and the mass of the proton, m_p. The natural classification is then given in table. Another important property is their range: the distance to which the interaction can be felt, and the type of quantity they couple to. Let me look a little closer at each of these in turn.

Table: A summary of the four fundamental forces

Force	Range	Strength	Acts on
Gravity	∞	$G_N \approx 6\ 10^{-39}$	All particles (mass and energy)
Weak Force	$< 10^{-18}$m	$G_F \approx 1\ 10^{-5}$	Leptons, Hadrons
Electromagnetism	∞	1/137	All charged particles
Strong Force	$\approx 10^{-15}$m	$g^2 \approx 1$	Hadrons

In order to set the scale we need to express everything in a natural set of units. Three scales are provided by $\hbar$ and c and e - actually one usually works in units where these two quantities are 1 in high energy physics. For the scale of mass we use the mass of the proton. In summary (for $e = 1$ we use electron volt as natural unit of energy)

$$\hbar = 6.58\times10^{-22} \text{ MeV s}$$

$\hbar c = 1.97 \times 10^{-13}$ MeV

$m_p = 938$ MeV/c^2

GRAVITY

The theory of gravity can be looked at in two ways: The old fashioned Newtonian gravity, where the potential is proportional to the rest mass of the particles,

$$V = \frac{G_N m_1 m_2}{r}$$

We find that $G_N m_p^2/\hbar c$ is dimensionless, and takes on the value

$G_N m_p^2/\hbar c = 5.9046486 \times 10^{-39}$

There are two more levels to look at gravity. One of those is Einstein's theory of gravity, which in the low-energy small-mass limit reduces to Newton's theory. This is still a *classical* theory, of a classical gravitational field.

The quantum theory, where we reexpress the field in their quanta has proven to be a very tough stumbling block - When one tries to generalise the approach taken for QED, every expression is infinite, and one needs to define an infinite number of different infinite constants. This is not deemed to be acceptable - i.e., it doesn't define a theory. Such a model is called unrenormalisable. We may return to the problem of quantum gravity later, time permitting.

ELECTROMAGNETISM

Electro-magnetism, i.e., QED, has been discussed in some detail in the previous chapter. Look there for a discussion. The coupling constant for the theory is

$$\alpha = \frac{e^2}{4\pi\varepsilon_0 \hbar c}$$

WEAK FORCE

This manifests itself through nuclear β decay,

$$\vec{n}p + e^- + \overline{\nu_e}$$

The standard coupling for this theory is called the Fermi coupling, G_F, after its discoverer. After the theory was introduced it was learned that there were physical particles that mediate the weak force, the $W^{\pm}$ and the Z^0 bosons. These are very heavy particles (their mass is about 80 times the proton mass!), which is why they have such a small range - fluctuations where I need to create that much mass are rare. The $W^{\pm}$ bosons are charged, and the Z^0 boson is neutral. The typical β decay referred to above is mediated by a W^- boson as can be seen in the Feynman diagram figure. The reason for this choice is that it conserves charge at each point (the charge of a proton and a W^- is zero, the charge of an electron and a neutrino is -1, the same as that of a W^-).

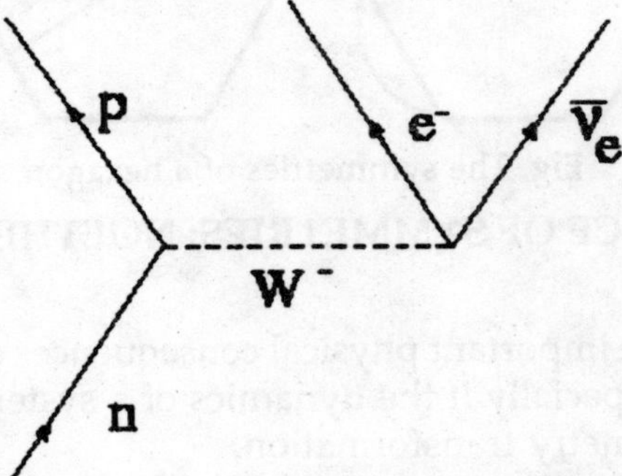

Fig. The Feynman diagram for the weak decay of a neutron.

STRONG FORCE

The strong force is what keeps nuclei together. It is described by a theory called QCD, which described the forces between fermions called quarks that make up the hadrons. These forces are mediated by spin-1 bosons called gluons. Notice that this is a case where a series in powers of the coupling constant does not make a lot of sense, since higher powers have about the same value as lower powers. Such a theory is called non-perturbative.

SYMMETRIES AND PARTICLE PHYSICS

Symmetries in physics provide a great fascination to us - one of the hang-ups of mankind. We can recognise a symmetry easily, and they provide a great tool to classify shapes and patterns. There is an important area of mathematics called group theory, where one studies the transformations under which an object is symmetric. In order to make this statement seem less abstract, let me look at a simple example, a regular hexagon in a plane. As can be seen in figure this object is symmetric (i.e., we can't distinguish the new from the old object) under rotations around centre over angles of a multiple of 60°, and under reflection in any of the six axes sketched in the second part of the figure.

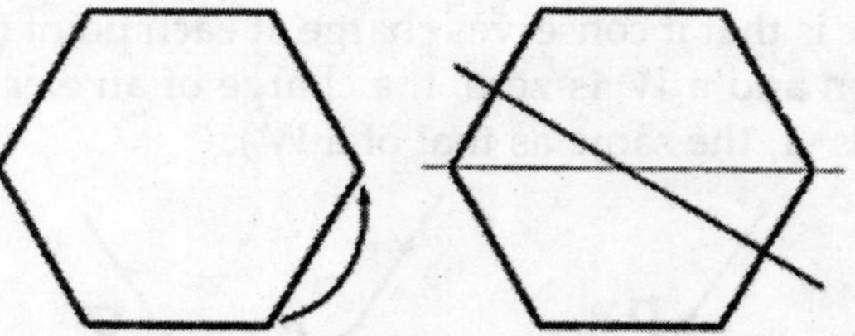

Fig. The symmetries of a hexagon

IMPORTANCE OF SYMMETRIES: NOETHER'S THEOREM

There are important physical consequences of symmetries in physics, especially if the dynamics of a system is invariant under a symmetry transformation.

There is a theorem, due to Emily Noether, one of the most important (female) mathematicians of this century, that states that for any *continuous* symmetry there is a conserved quantity.

So what is a continuous symmetry? Think about something like spherical symmetry - a sphere is invariant under any rotation about its centre, no matter what the rotation angle. The continuity of choice of parameter in a transformation is what makes the set of transformations continuous. Another way of saying the same thing is that the transformation can be arbitrarily close to the unit transformation, i.e., it can do almost nothing at all.

LORENZ AND POINCARÉ INVARIANCE

One of the most common continuous symmetries of a relativistic theory is Lorentz invariance, i.e., the dynamics is the same in any Lorentz frame. The group of Lorentz transformations can be decomposed into two parts:

- Boosts, where we go from one Lorentz frame to another, i.e., we change the velocity.
- Rotations, where we change the orientation of the coordinate frame.

There is a slightly larger group of symmetries, called the Poincaré group, obtained when we add translations to the set of symmetries - clearly the dynamics doesn't care where we put the orbit of space.

The set of conserved quantities associated with this group is large. Translational and boost invariance implies conservation of four momentum, and rotational invariance implies conservation of angular momentum.

INTERNAL AND SPACE-TIME SYMMETRIES

Above I have mentioned angular momentum, the vector product of position and momentum. This is defined in terms of properties of space (or to be more generous, of space-time). But we know that many particles carry the spin of the particle to form the total angular momentum,

J = L + S.

The invariance of the dynamics is such that J is the conserved quantity, which means that we should not just rotate in ordinary space, but in the abstract "intrinsic space" where S is defined. This is something that will occur several times again, where a symmetry has a combination of a space-time and intrinsic part.

DISCRETE SYMMETRIES

Let us first look at the key discrete symmetries - parity *P* (space inversion) charge conjugation *C* and time-reversal *T*.

PARITY *P*

Parity is the transformation where we reflect each point in the origin, $x \rightarrow - x$. This transformation should be familiar to you. Let us think of the one dimensional harmonic oscillator, with Hamiltonian

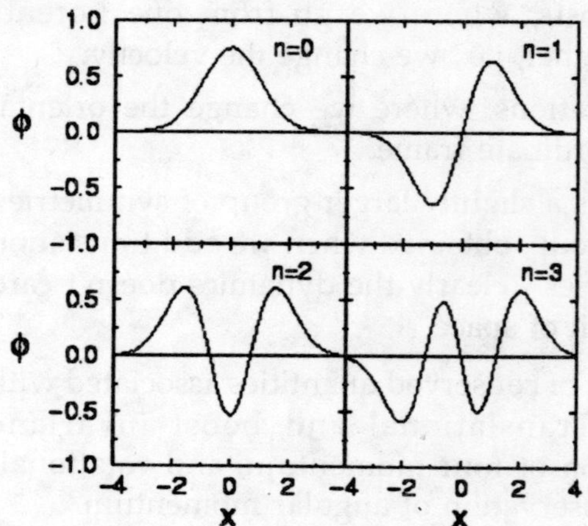

Fig. The first four harmonic oscillator wave functions

$$\frac{\hbar^2}{2m}\frac{d}{dx^2}+\frac{1}{2}m\omega^2x^2.$$

The Hamiltonian does not change under the substitution $x \rightarrow - x$. The well-known eigenstates to this problem are either even or odd under this transformation, see Fig. and thus have either even or odd parity,

$$P\psi\,(x, t) = \psi\,(- x, t) = \pm\psi\,(x, t),$$

where P is the transformation that take $x \rightarrow - x$. For

$$P\psi\,(x, t) = \psi\,(x, t)$$

we say that the state has even parity, for the minus sign we speak about negative parity. These are the only two allowed eigenvalues, as can be seen from looking at the probability density $|\,\psi\,(x, y)\,|^2$. Since this must be *invariant*, we find that

$$|\,\psi\,(x, y)\,|^2 = |\,P\psi\,(x, y)\,|^2$$

which shows that the only *real* eigenvalues for P are ±1. One can show that there is a relation between parity and the orbital angular momentum quantum number L, $\pi = (-1)^L$, which relates two space-time symmetries.

It is found, however, that parity also has an intrinsic part, which is associated with each type of particle. A photon (γ) has negative parity. This can be understood from the following classical analogy. When we look at Maxwell's equation for the electric field,

$$\nabla E(x,t) = \frac{1}{\varepsilon_0 \rho}(x,t),$$

we find that upon releversal of the coordinates this equation becomes

$$\nabla E(-x,t) = \frac{1}{\varepsilon_0 \rho}(-x,t),$$

The additional minus sign, which originates in the change of sign of ∇ is what gives the electric field and thus the photon its negative intrinsic parity.

We shall also wish to understand the parity· of particles and antiparticles. For fermions (electrons, protons,...) we have the interesting relation $P_f P \bar{f}$ = - 1, which will come in handy later!

CHARGE CONJUGATION C

The name of this symmetry is somewhat of a misnomer. Originally it stems from QED, where it was found that a set of interacting electrons behaves exactly the same way as a similar set of positrons. So if we change the sign of all charges the dynamics is the same. Actually, the symmetry generalises a little bit, and in general refers to a transformation where we change all particles in their antiparticles.

Once again we find $C^2 = 1$, and the only possible eigenvalues of this symmetry are ±1. An uncharged particle like the photon that is its own antiparticle, must be an eigenstate of the symmetry operation, and it is found that it has eigenvalue -1,

$$C\psi_\gamma = -\psi_\gamma.$$

(Here ψ_γ is the wave function of the photon.) This can be shown from Maxwell's equation as before, since ρ chanegs sign under charge conjugation.

For a combination of a particle and an antiparticle, we find that $C_f C \overline{f}$ = - 1 for fermions, and +1 for bosons.

TIME REVERSAL *T*

On a microscopic scale it is not very apparent whether time runs forward or backwards, the dynamics where we just change the sign of time is equally valid as the original one. This corresponds to flipping the sign of all momenta in a Feynman diagram, so that incoming particles become outgoing particles and vice-versa. This symmetry is slightly nastier, and acts on both space-time and intrinsic quantities such as spin in a complicated way. The space time part is found to be

$$T\,\psi(\mathrm{r}, t) = \psi^{*}(\mathrm{r}, -t).$$

Combined with its intrinsic part we find that it has eigenvalues $\pm i$ for fermions (electrons, etc.) and ±1 for bosons (photons, etc.).

THE *CPT* THEOREM

A little thought shows that all three symmetries mentioned above appear very natural - but that is a theorist's argument. The real key test is experiment, not a theorist's nice ideas! In 1956 C.N. Yang and T.D. Lee analysed the experimental evidence for these symmetries. They realised there was good evidence of these symmetries in QED and QCD (the theory of strong interactions). There was no evidence that parity was a symmetry of the weak interactions - which was true, since it was shown soon thereafter that these symmetries are broken, in a beautiful experiment led by "Madame" C.S. Wu.

There is a fairly strong proof that only minimal physical assumptions (locality, causality) that the product of *C*, *P* and *T* is a good symmetry of any theory. Up to now experiment has not shown any breaking of this product. We would have to rethink a lot of basic physics if this symmetry is not present. I am reasonably confident that if breaking is ever found there will be ten models that can describe it within a month!

CP VIOLATION

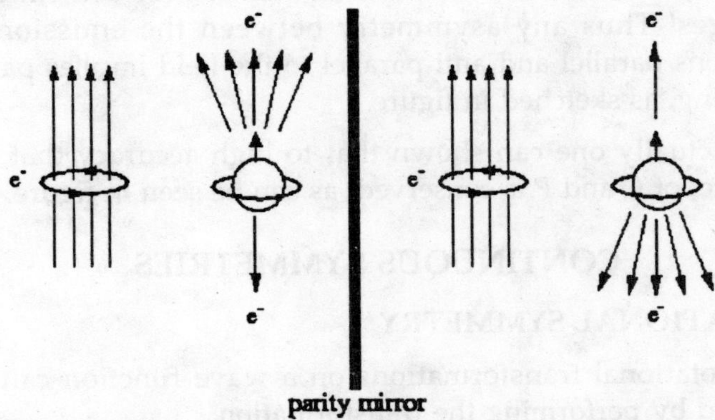

Figure: Parity breaking for the β decay of ^{60}Co

The first experimental confirmation of symmetry breaking was found when studying β^- the decay of ^{60}Co,

$$^{60}Co \rightarrow {}^{60}Ni + e^- + \overline{\nu_e}\,.$$

This nucleus has a ground state with non-zero spin, which can be oriented in a magnetic field.

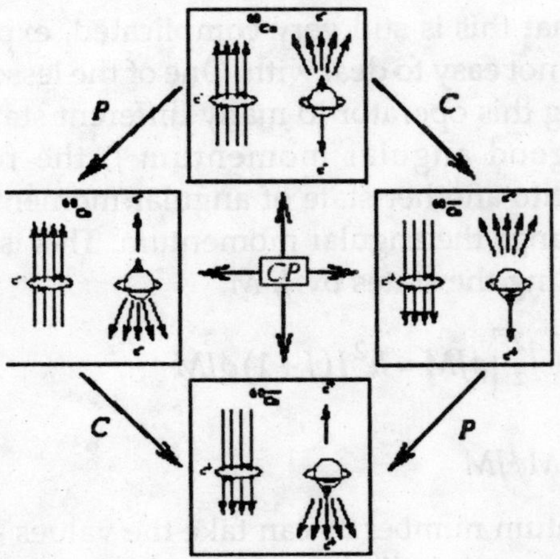

Fig. *CP* symmetry for the decay of ^{60}Co

A magnetic field is a pseudo-vector, which means that under parity it goes over into itself B $\rightarrow$ B. So does the spin of the nucleus, and we thus have established that under parity the situation under which the nucleus emits electrons should be invariant. But the direction in which they are emitted changes! Thus any asymmetry between the emission of electrons parallel and anti-parallel to the field implies parity breaking, as sketched in figure.

Actually one can shown that to high accuracy that the product of *C* and *P* is conserved, as can be seen in figure.

CONTINUOUS SYMMETRIES

ROTATIONAL SYMMETRY

Rotational transformations on a wave function can be applied by performing the transformation

$$\exp\left[i\left(\theta_x \hat{J}_x + \theta_y \hat{J}_y + \theta_z \hat{J}_z\right)\right]$$

on a wave function. This is slightly simpler for a particle without spin, since we shall only have to consider the orbital angular momentum,

$$\hat{L} = \hat{p} \times r = i\hbar r \times \nabla.$$

Notice that this is still very complicated, exponentials of operators are not easy to deal with. One of the lessons we learn from applying this operator to many different states, is that if a state has good angular momentum *J*, the rotation can transform it into another state of angular momentum *J*, but it will never change the angular momentum. This is most easily seen by labelling the states by *J*, *M*:

$$\left[\hat{J}_x^2 + \hat{J}_y^2 + \hat{J}_z^2\right]\phi JM = \hbar^2 J(J+1)\phi JM$$

$$\hat{J}_z^2 \phi JM = M\phi JM$$

The quantum number *M* can take the values - *J*, - *J* + 1,..., *J* - 1, *J*, so that we typically have 2*J* + 1 components for each *J*.

The effect of the exponential transformation on a linear combination of states of identical J is to perform a linear transformation between these components. I shall show in a minute that such transformation can be implemented by unitary matrices. The transformations that implement these transformations are said to correspond to an irreducible representation of the rotation group (often denoted by SO(3)).

Let us look at the simplest example, for spin 1/2. We have two states, one with spin up and one with spin down, $\psi_\pm$. If the initial state is $\psi = \alpha_+\psi_+ + \alpha-\psi-$, the effect of a rotation can only be to turn this into $\psi' = \alpha'_+\psi'_+ + \alpha'-\psi'-$. Since the transformation is linear (if I rotate the sum of two objects, I might as well rotate both of them) we find

$$\begin{pmatrix} \alpha'_+ \\ \alpha'_- \end{pmatrix} = \begin{pmatrix} U_{++} & U_{-+} \\ U_{+-} & U_{--} \end{pmatrix} \begin{pmatrix} \alpha'_+ \\ \alpha'_- \end{pmatrix}$$

Since the transformation can not change the length of the vector, we must have $\int|\psi'|^2 = 1$. Assuming $\int|\psi_\pm|^2 = 1, \int\psi^\bullet + \psi- = 0$ we find

$$U^t U = 1$$

with

$$U^t = \begin{pmatrix} U^\bullet_{++} & U^\bullet_{+-} \\ U^\bullet_{-+} & U^\bullet_{--} \end{pmatrix}$$

the so-called hermitian conjugate.

We can write down matrices that in the space of $S = 1/2$ states behave the same as the angular momentum operators. These are *half* the well known Pauli matrices

$$\sigma_x = \begin{pmatrix} 0 & 1 \\ 1 & 0 \end{pmatrix}, \sigma_y = \begin{pmatrix} 0 & -i \\ i & 0 \end{pmatrix}, \sigma_z = \begin{pmatrix} 1 & 0 \\ 0 & -1 \end{pmatrix}$$

and thus we find that

$$U(\theta) = \exp\left[i(\theta x\sigma x/2 + \theta y\sigma y/2 + \theta z\sigma z/2)\right].$$

I don't really want to discuss how to evaluate the exponent of a matrix, apart from one special case. Suppose we perform a 2π rotation around the z axis, $\theta = (0, 0, 2\pi)$. We find

$$U(0, 0, 2\pi) = \exp[i\pi\begin{pmatrix}1 & 0\\ 0 & -1\end{pmatrix}].$$

Since this matrix is diagonal, we just have to evaluate the exponents for each of the entries (this corresponds to using the Taylor series of the exponential),

$$U(0, 0, 2\pi) = \begin{pmatrix}\exp[i\pi] & 0\\ 0 & \exp[-i\pi]\end{pmatrix}$$

$$= \begin{pmatrix}-1 & 0\\ 0 & -1\end{pmatrix}.$$

To our surprise this does not take me back to where I started from. Let me make a small demonstration to show what this means.........

Finally what happens if we combine states from two irreducible representations? Let me analyse this for two spin 1/2 states,

$$\psi = (\alpha_+^1\psi_+^1 + \alpha_-^1\psi_-^1)(\alpha_+^2\psi_+^2 + \alpha_-^2\psi_-^2)$$

$$= \alpha_+^1\alpha_+^2\psi_+^1\psi_+^2 + \alpha_+^1\alpha_-^2\psi_+^1\psi_-^2 + \alpha_-^1\alpha_+^2\psi_-^1\psi_+^2\alpha_-^1\alpha_-^2\psi_-^1\psi_-^2.$$

The first and the last product of ψ states have an angular momentum component ±1 in the z direction, and must does at least have $J = 1$. The middle two combinations with both have $M = M_1 + M_2 = 0$ can be shown to be a combination of a $J = 1$, $M = 0$ and a $J = 0$, $M = 0$ state. Specifically,

$$\frac{1}{\sqrt{2}}[\psi_+^1\psi_-^2 \quad \psi_-^1\psi_+^2]$$

transforms as a scalar, it goes over into itself. the way to see that is to use the fact that these states transform with the same U, and substitute these matrices. The result is proportional to where we started from. Notice that the triplet

(S = 1) is symmetric under interchange of the two particles, whereas the singlet (S = 0) is antisymmetric. This relation between symmetry can be exhibited where the horizontal direction denotes symmetry, and the vertical direction denotes antisymmetry. This technique works for all unitary groups.

Fig. The Young tableau for the multiplication 1/2×1/2 = 0 + 1.

The coupling of angular momenta is normally performed through Clebsch-Gordan coefficients, as denoted by

$$\langle j_1 m_1\, j_2 m_2 | JM \rangle.$$

We know that $M = m_1 + m_2$. Further analysis shows that J can take on all values $|j_1 - j_2|, |j_1 - j_2| + 1, |j_1 - j_2| + 2, j1 + j2$.

SYMMETRIES AND SELECTION RULES

We shall often use the exact symmetries discussed up till now to determine what is and isn't allowed. Let us, for instance, look at

REPRESENTATIONS OF SU(3) AND MULTIPLICATION RULES

A very important group is SU(3), since it is related to the colour carried by the quarks, the basic building blocks of QCD.

The transformations within SU(3) are all those amongst a vector consisting of three complex objects that conserve the length of the vector. These are all three-by-three unitary matrices, which act on the complex vector ψ by

$$\psi \to U\psi$$

$$= \begin{pmatrix} U_{11} & U_{12} & U_{13} \\ U_{21} & U_{22} & U_{23} \\ U_{31} & U_{32} & U_{33} \end{pmatrix} \begin{pmatrix} \psi_1 \\ \psi_2 \\ \psi_3 \end{pmatrix}$$

The complex conjugate vector can be shown to transform as

$$\psi^{\bullet} \to \psi^{\bullet} U^{t},$$

with the inverse of the matrix. Clearly the fundamental representation of the group, where the matrices representing the transformation are just the matrix transformations, the vectors have length 3. The representation is usually labelled by its number of basis elements as 3. The one the transforms under the inverse matrices is usually denoted by $\overline{3}$.

What happens if we combine two of these objects, ψ and $X^{\bullet}$? It is easy to see that the inner product of ψ and $X^{\bullet}$ is scalar,

$$X^{\bullet}.\psi \rightarrow X^{\bullet}\ UtU\psi = X^{\bullet}.\psi$$

where we have used the unitary properties of the matrices the remaining 8 components can all be shown to transform amongst themselves, and we write

$$3 \otimes \overline{3} = 1 \otimes 8.$$

Of further interest is the product of three of these vectors,

$$3 \otimes 3 \otimes 3 = 1 \otimes 8 \otimes 8 \otimes 10.$$

BROKEN SYMMETRIES

Of course one cannot propose a symmetry, discover that it is not realised in nature ("the symmetry is broken"), and expect that we learn something from that about the physics that is going on. But parity is broken, and we still find it a useful symmetry! That has to do with the manner in which it is broken, only weak interactions - the exchange of $W^{\pm}$ and Z bosons - break them. Any process mediated by strong, electromagnetic or (probably) gravitational forces conserves the symmetry. This is one example of a symmetry that is only mildly broken, i.e., where the conserved quantities are still recognisable, even though they are not exactly conserved.

In modern particle physics the way symmetries are broken teaches us a lot about the underlying physics, and it is one of the goals of grand-unified theories (GUTs) to try and understand this.

GAUGE SYMMETRIES

One of the things which needs to be mentioned, is of a certain class of local symmetries (i.e., symmetries of the theory at each point in space and time) called gauge symmetries. This is a key idea in almost all modern particle physics theories, so much so that they are usually labelled by the local symmetry group. Local symmetries are not directly observable, and do not have immediate consequences. They allow for a mathematically consistent and simple formulation of the theories, and in the end predict the particle that are exchanged - the gauge particles, as summarised in table.

Table: The four fundamental forces and their gauge particles

Gravitation	**graviton(?)**
QED	photon
Weak	$W^{\pm}$, Z^0
Strong	gluons

SYMMETRIES OF THE THEORY OF STRONG INTERACTIONS

The first time people realised the key role of symmetries was in the plethora of particles discovered using the first accelerators. Many of those were composite particle (to be explained later) bound by the strong interaction.

THE FIRST SYMMETRY: ISOSPIN

The first particles that show an interesting symmetry are actually the nucleon and the proton. Their masses are remarkably close,

$$M_{\mathrm{p}} = 939.566\ MeV/c^2 \quad M_{\mathrm{n}} = 938.272\ MeV/c^2.$$

If we assume that these masses are generated by the strong interaction there is more than a hint of symmetry here. Further indications come from the pions: they come in three charge states, and once again their masses are remarkably similar,

$$M^{\pi^+} = M^{\pi^-} = 139.567\ MeV/c^2,\ M^{\pi^0} = 134.974\ MeV/c^2.$$

This symmetry is reinforced by the discovery that the interactions between nucleon (p and n) is independent of charge, they only depend on the nucleon character of these particles - the strong interactions see only one nucleon and one pion. Clearly a continuous transformation between the nucleons and between the pions is a symmetry. The symmetry that was proposed (by Wigner) is an internal symmetry like spin symmetry called isotopic spin or isospin. It is an abstract rotation in isotopic space, and leads to similar type of states with isotopic spin $I = 1/2, 1, 3/2, \ldots$. One can define the third component of isospin as

$$Q = e(I_3 + B),$$

where B is the baryon number ($B = 1$ for n, p, 0 for π). We thus find

	B	Q/e	I	I_3
n	1	0	1/2	-1/2
p	1	1	1/2	1/2
π^-	0	-1	1	-1
π^0	0	0	1	0
π^+	0	1	1	1

Notice that the energy levels of these particles are split by a magnetic force, as ordinary spins split under a magnetic force.

STRANGE PARTICLES

The British physicists Rochester and Butler (from across the street) observed new particles in cosmic ray events. (Cosmic rays where the tool before accelerators existed - they are still used due to the unbelievably violent processes taking place in the cosmos. We just can't produce particles like that in the lab. (Un)fortunately the number of highly energetic particles is very low, and we won't see many events.) These particles came in two forms: a neutral one that decayed into a π^+ and a π^-, and a positively charge one that decayed into a μ^+(heavy electron) and a photon, as sketched in figure.

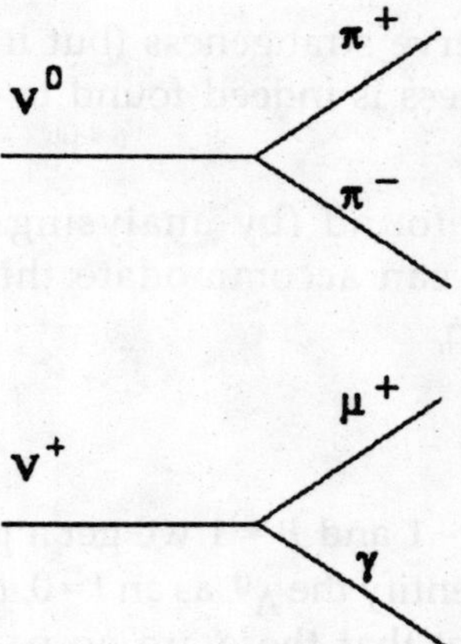

Fig. The decay of *V* particles

The big surprise about these particles was how long they lived. There are many decay time scales, but typically the decay times due to strong interactions are very fast, of the order of a femto second (10^{-15} s). The decay time of the *K* mesons was about 10^{-10} s, much more typical of a weak decay. Many similar particles have since been found, both of mesonic and baryonic type (like pions or like nucleons). These are collectively know as strange particles. Actually, using accelerators it was found that strange particles are typically formed in pairs, e.g.,

$$\pi^+ + p \rightarrow \underbrace{\Lambda^0}_{\text{baryon}} + \underbrace{K^0}_{\text{meson}}$$

This mechanism was called associated production, and is highly suggestive of an additive conserved quantity, such as charge, called strangeness. If we assume that the has strangeness -1, and the K_0 +1, this balances

$\pi^+ + p \rightarrow \Lambda^0 + K^0$

0 + 0 - = -1 + 1

The weak decay

+ *p*

-1 0 + 0,

does not conserve strangeness (but it conserves baryon number). This process is indeed found to take much longer, about 10^{-10} s.

Actually it is found (by analysing many resonance particles) that we can accommodate this quantity in our definition of isospin,

$$Q = e(I_3 + \frac{B+S}{2})$$

Clearly for $S = -1$ and $B = 1$ we get a particle with $I_3 = 0$. This allows us to identify the Λ^0 as an $I = 0$, $I_{3=0}$ particle, which agrees with the fact that there are no particles of different charge and a similar mas: and strong interaction properties.

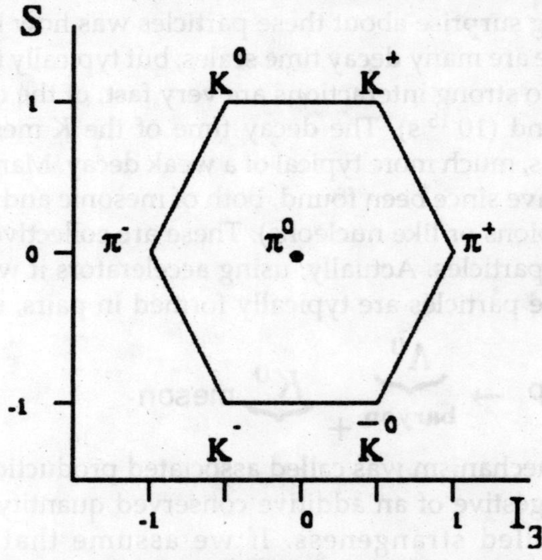

Fig. A possible arrangement for the states of the septet

The kaons come in three charge states $K^{\pm}$, K^0 with masses $m_{K^{\pm}} = 494$ *MeV*, $m_{K^0} = 498$ *MeV*. In similarity with pions, which form an $I = 1$ multiplet, we would like to assume a $I = 1$ multiplet of K's as well. This is problematic since we have to assume $S = 1$ for all these particles: we cannot satisfy

$$Q = e(I_3 + \frac{1}{2})$$

for isospin 1 particles. The other possibility $I = 3/2$ doesn't fit with only three particles. Further analysis shows that the the K^+ is the antiparticle of K^-, but K^0 is not its own antiparticle (which is true for the pions. So we need four particles, and the assignments are $S = 1, I = 1/2$ for K^0 and K^-, $S = -1, I = 1/2$ for K^+ and $\overline{K}^0$. Actually, we now realise that we can summarise all the information about K's and π's in one multiplet, suggestive of a (pretty badly broken!) symmetry.

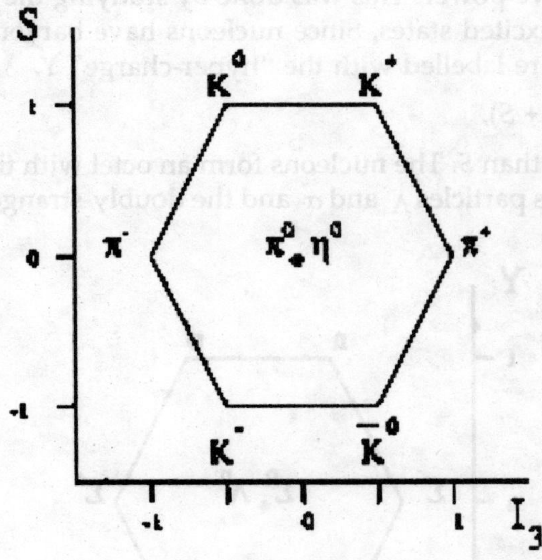

Fig. Octet of mesons

However, it is hard to find a sensible symmetry that gives a 7-dimensional multiplet. It was argued by Gell-Mann and Ne'eman in 1961 that a natural extension of isospin symmetry would be an SU(3) symmetry. We have argued before that one of the simplest representations of SU(3) is 8 dimensional symmetry. A mathematical analysis shows that what is missing is a particle with $I = I_3 = S = 0$. Such a particle is known, and is called the η^0. The breaking of the symmetry can be seen from the following mass table:

$$m_{\pi^\pm} = 139 \ MeV$$

$$m_{\pi^0} = 134 \ MeV$$

$m_{K^{\pm}} = 494\ MeV$

$m_{\overset{(-)}{k}{}^0} = 498\ MeV$

$m_{\eta^0} = 549\ MeV$

The resulting multiplet is often represented like in figure.

In order to have the scheme make sense we need to show its predictive power. This was done by studying the nucleons and their excited states. Since nucleons have baryon number one, they are labelled with the "hyper-charge" Y,

$Y = (B + S)$,

rather than S. The nucleons form an octet with the single-strangeness particles Λ and σ and the doubly-strange cascade particle Ξ,.

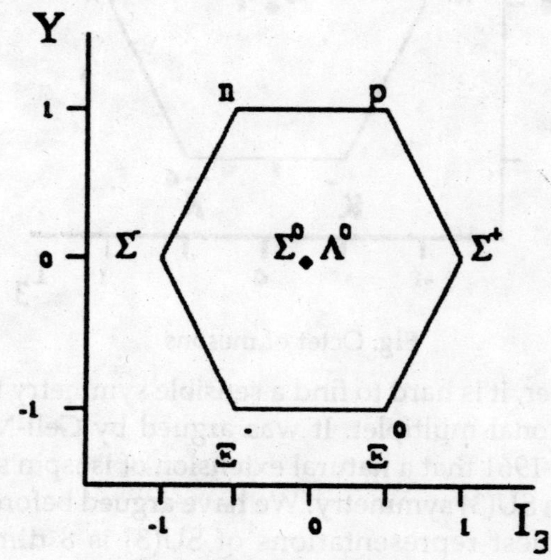

Fig. Octet of nucleons

The masses are

$M_n = 938\ MeV$

$M_p = 939\ MeV$

$M_{\Lambda^0} = 1115\ MeV$

$M^{\Sigma^+} \quad = \quad 1189 \; MeV$

$M^{\Sigma^0} \quad = \quad 1193 \; MeV$

$M^{\Sigma^-} \quad = \quad 1197 \; MeV$

$M^{\Xi^0} \quad = \quad 1315 \; MeV$

$M^{\Xi^-} \quad = \quad 1321 \; MeV$

All these particles were known before the idea of this symmetry. The first confirmation came when studying the excited states of the nucleon. Nine states were easily incorporated in a decuplet, and the tenth state (the Ω^-, with strangeness -3) was predicted. It was found soon afterwards at the predicted value of the mass.

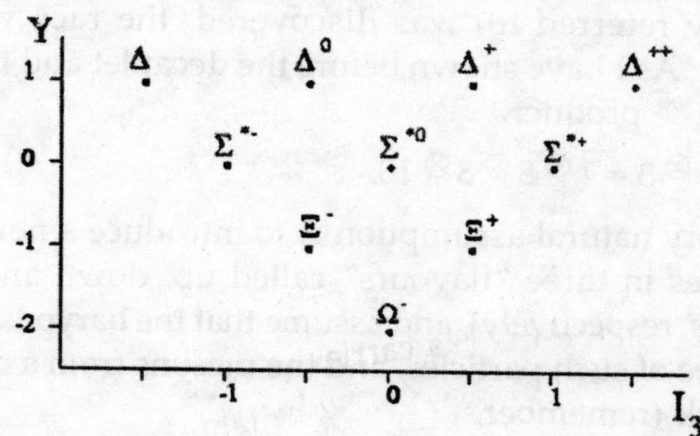

Fig. Decuplet of excited nucleons

The masses are

$M\Delta \quad = \quad 1232 \; MeV$

$M\Sigma^\bullet \quad = \quad 1385 \; MeV$

$M\Xi^\bullet \quad = \quad 1530 \; MeV$

$M\Omega \quad = \quad 1672 \; MeV$

(Notice almost that we can fit these masses as a linear function in Y, as can be seen in figure. This was of great help in finding the Ω.)

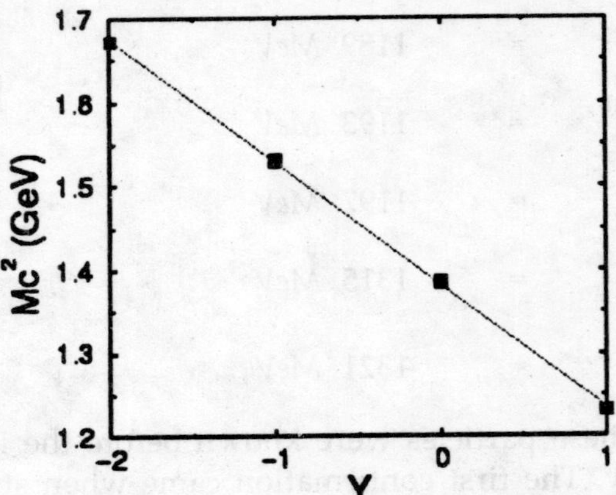

Fig. A linear fit to the mass of the decuplet

THE QUARK MODEL OF STRONG INTERACTIONS

Once the eightfold way (as the SU(3) symmetry was poetically referred to) was discovered, the race was on to explain it. As I have shown before the decaplet and two octets occur in the product

$$3 \otimes 3 \otimes 3 = 1 \otimes 8 \otimes 8 \otimes 10.$$

A very natural assumption is to introduce a new particle that comes in three "flavours" called up, down and strange (*u*, *d* and *s*, respectively), and assume that the baryons are made from three of such particles, and the mesons from a quark and anti-quark (remember,

$3 \otimes \bar{3} = 1 \otimes 8$.)

Each of these quarks carries one third a unit of baryon number. The properties can now be tabulated,

Table: The properties of the three quarks.

Quark	label	spin	Q/e	I	I_3	S	B
Up	u	$\frac{1}{2}$	$+\frac{2}{3}$	$\frac{1}{2}$	$+\frac{1}{2}$	0	$\frac{1}{3}$
Down	d	$\frac{1}{2}$	$-\frac{1}{3}$	$\frac{1}{2}$	$-\frac{1}{2}$	0	$\frac{1}{3}$

Strange	s	$\frac{1}{2}$	$-\frac{1}{3}$	0	0	-1	$\frac{1}{3}$

In the multiplet language I used before, we find that the quarks form a triangle, as given in Fig..

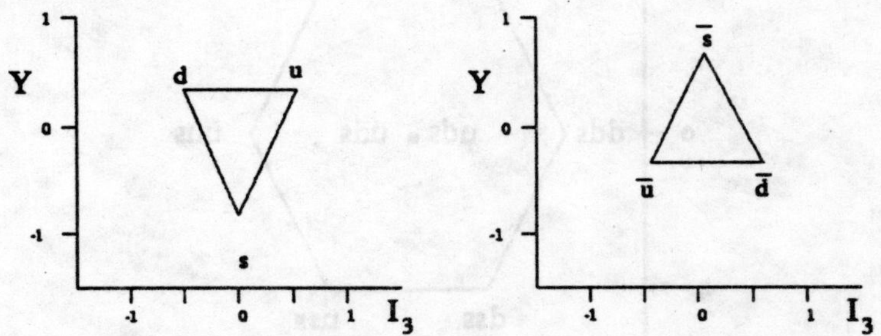

Fig. The multiplet structure of quarks and antiquarks

Once we have made this assigment, we can try to derive what combination corresponds to the assignments of the meson octet, figure. We just make all possible combinations of a quark and antiquark, apart from the scalar one $\eta' = u\bar{\mu} + d\bar{d} + c\bar{c}$ (why?).

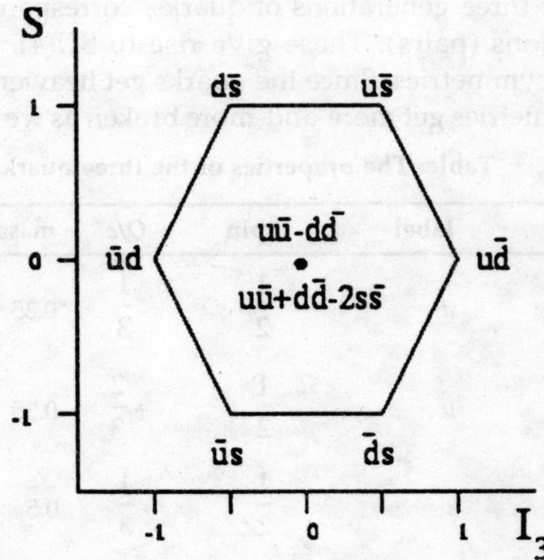

Figure: quark assignment of the meson octet

A similar assignment can be made for the nucleon octet, and the nucleon decaplet,

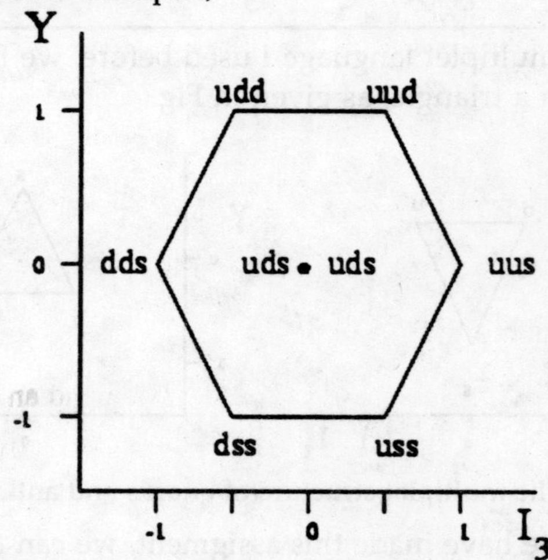

Fig. Quark assignment of the nucleon octet

SU(4),...

Once we have three flavours of quarks, we can ask the question whether more flavours exists. At the moment we know of three generations of quarks, corresponding to three generations (pairs). These give rise to SU(4), SU(5), SU(6) flavour symmetries. Since the quarks get heavier and heavier, the symmetries get more-and-more broken as we add flavours.

Table: The properties of the three quarks.

Quark	label	spin	Q/e	mass (GEV/c^2)
Down	*d*	$\frac{1}{2}$	$-\frac{1}{3}$	0.35
Up	*u*	$\frac{1}{2}$	$+\frac{2}{3}$	0.35
Strange	*s*	$\frac{1}{2}$	$-\frac{1}{3}$	0.5
Charm	*c*	$\frac{1}{2}$	$+\frac{2}{3}$	1.5

Bottom	b	$\frac{1}{2}$	$-\frac{1}{3}$	4.5
Top	t	$\frac{1}{2}$	$+\frac{2}{3}$	93

COLOUR SYMMETRY

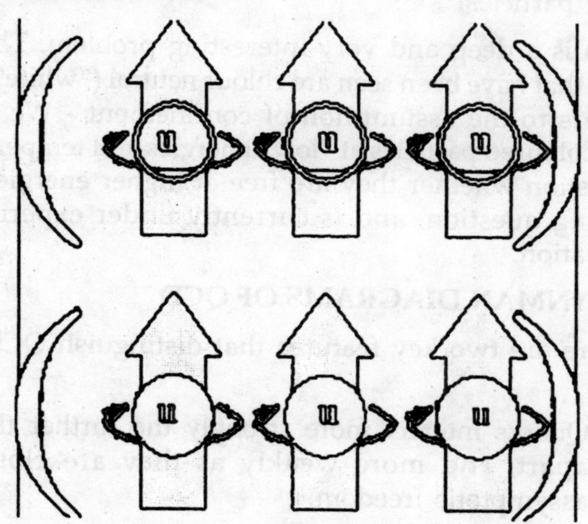

Fig. **The Δ^{++} in the quark model.**

So why don't we see fractional charges in nature? This is an important point! In so-called deep inelastic scattering we see pips inside the nucleon - these have been identified as the quarks. We do not see any direct signature of individual quarks. Furthermore, if quarks are fermions, as they are spin 1/2 particles, what about antisymmetry of their wavefunction? Let us investigate the Δ^{++}, see Fig. which consists of three u quarks with identical spin and flavour (isospin) and *symmetric* spatial wavefunction,

$$\psi_{total} = \psi_{space} \times \psi_{spin} \times \psi_{flavor},$$

This would be symmetric under interchange, which is unacceptable. Actually there is a simple solution. We "just" assume that there is an additional quantity called colour, and take the colour wave function to be antisymmetric:

$$\psi_{total} = \psi_{space} \times \psi_{spin} \times \psi_{flavor} \times \psi_{color}$$

We assume that quarks come in three colours. This naturally leads to yet another $SU(3)$ symmetry, which is actually related to the gauge symmetry of strong interactions, QCD. So we have shifted the question to: why can't we see coloured particles?

This is a deep and very interesting problem. The only particles that have been seen are colour neutral ("white") ones. This leads to the assumption of confinement - We cannot liberate coloured particles at "low" energies and temperatures! The question whether they are free at higher energies is an interesting question, and is currently under experimental consideration.

THE FEYNMAN DIAGRAMS OF QCD

There are two key features that distinguish QCD from QED:

1. Quarks interact more strongly the further they are apart, and more weakly as they are close by - assymptotic freedom.
2. Gluons interact with themselves

The first point can only be found through detailed mathematical analysis. It means that free quarks can't be seen, but at high energies quarks look more and more like free particles. The second statement make QCD so hard to solve. The gluon comes in 8 colour combinations (since it carries a colour and anti-colour index, minus the scalar combination). The relevant diagrams are sketches in Figure. Try to work out yourself how we satisfy colour charge conservation!

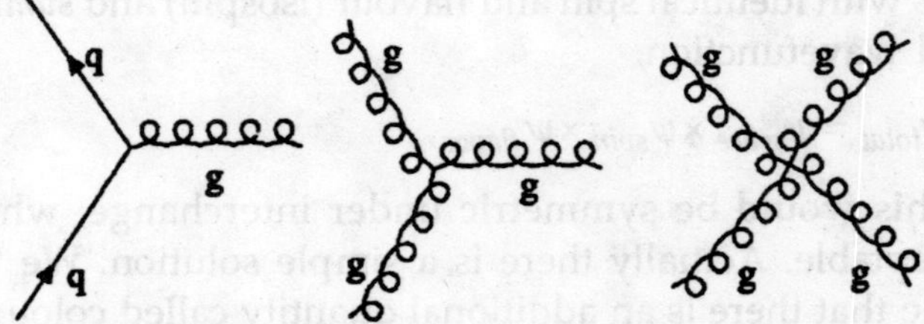

Fig. The basic building blocks for QCD feynman diagrams

JETS AND QCD

One way to see quarks is to use the fact that we can liberate quarks for a short time, at high energy scales. One such process is $e^+e^- \rightarrow q$, which use the fact that a photon can couple directly to $q\bar{q}$. The quarks don't live very long and decay by producing a "jet" a shower of particles that results from the deacay of the quarks. These are all "hadrons", mesons and baryons, since they must couple through the strong interaction. By determining the energy in each if the two jets we can discover the energy of the initial quarks, and see whether QCD makes sense.

RELATIVISTIC KINEMATICS

One of the features of particle physics is the importance of special relativity. This occurs at a very fundamental level, since particle physics is all about creating and annihilating particles. This can only occur if we can convert mass to energy and vice-versa. Thus Einstein's idea of the equivalnece between mass and energy plays an extremely fundamental rôle in this field of physics. In order for this to be possible we typically need processes that occur at velocities near the light velocity c, so that the kinematics (i.e., the description of momemnta and energy) of these processes requires relativity. In this chapter we shall succintly introduce the few necessary concepts - I hope that for most of you this is a review, but this chapter is intended to be self-contained and contains everything I shall need in relativistic kinematics.

LORENTZ TRANSFORMATIONS OF ENERGY AND MOMENTUM

As you may know, like we can combine position and time in one four-vector x = (x, ct), we can also combine energy and momentum in a single four-vector, p = (p, E/c). From the Lorentz transformation property of time and position, for a change of velocity along the x-axis from a coordinate system at rest to one that is moving with velocity v = (v_x, 0, 0) we have

$$x' = \gamma(v)(x - v/ct), \quad t' = \gamma(t - xvx/c^2),$$

we can derive that energy and momentum behave in the same way,

$$p'_x = \gamma'(v)(p_x - Ev/c^2) = mu_x\,\gamma(|\,u\,|),$$

$$E' = \gamma\,(v)(E - vp_x) = \gamma\,(|\,u\,|)m_0c^2.$$

To understand the context of these equations remember the definition of γ

$$\gamma(v) = 1/\sqrt{1-\beta^2}\,, \qquad \beta = \frac{v}{c}.$$

In Eq. we have also reexpressed the momentum energy in terms of a velocity u. This is measured relative to the rest system of a particle, the system where the three-momentum p = 0.

Now all these exercises would be interesting mathematics but rather futile if there was no further information. We know however that the full four-momentum is conserved, i.e., if we have two particles coming into a collision and two coming out, the sum of four-momenta before and after is equal,

$$E^{in}{}_1 + E^{in}{}_2 = \qquad E^{out}{}_1 + E^{out}{}_{2'}$$

$$\mathbf{p}^{in}{}_1 + \mathbf{p}^{in}{}_2 = \qquad \mathbf{p}^{out}{}_1 + \mathbf{p}^{out}{}_2.$$

INVARIANT MASS

One of the key numbers we can extract from mass and momentum is the *invariant mass*, a number independent of the Lorentz frame we are in

$$W^2c^4 = \left(\sum_i E_i\right)^2 - \left(\sum_i pi\right)^2 c^2.$$

This quantity takes it most transparent form in the centre-of-mass, where $\sum_i pi = 0$. In that case

$$W = E_{CM}/c^2,$$

and is thus, apart from the factor $1/c^2$, nothing but the energy in the CM frame. For a single particle $W = m_0$, the rest mass.

Most considerations about processes in high energy physics are greatly simplified by concentrating on the invariant mass. This removes the Lorentz-frame dependence of writing four momenta. I

As an exmaple we look at the collision of a proton and an antiproton at rest, where we produce two quanta of electromagnetic radiation (γ's), where the anti proton has three-momentum $(p, 0, 0)$, and the proton is at rest.

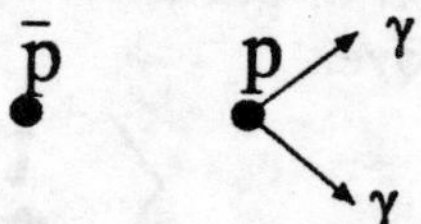

Fig. A sketch of a collision between a proton with velocity v and an antiproton at rest producing two *gamma* quanta.

The four-momenta are

$$p_p \quad = \quad (p_{lab}, 0, 0, \sqrt{m_p^2 c^4 + p^2 labc^2}\ )$$

$$p \quad = \quad (0, 0, 0, m_p c^2).$$

From this we find the invariant mass

$$W = \sqrt{2m_p^2 + 2m_p \sqrt{m_p^2 + p_{lab}^2 / c^2}}$$

If the initial momentum is much larger than m_p, more accurately

$$p_{lab} \gg m_p c,$$

we find that

$$W \approx \sqrt{2m_p plab / c},$$

which energy needs to be shared between the two photons, in equal parts. We could also have chosen to work in the CM frame, where the calculations get a lot easier.

TRANSFORMATIONS BETWEEN CM AND LAB FRAME

Even though the use of the invariant mass simplifies calculations considerably, it clearly does not provide all necessary information. It does suggest however, that a natural frame to analyse reactions is the CM frame. Often we shall analyse a process in this frame, and use a Lorentz transformation to get informations about processes in the laboratory frame. Since almost all processes involve the scattering (deflection) of one particle by another (or a number of others), this is natural example for such a procedure, see the sketch in Fig.. The same procedure can also be applied to the case of production of particles, such as the annihilation process discussed above.

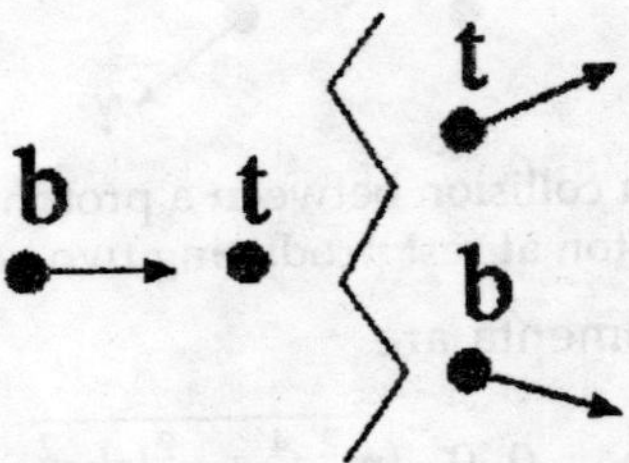

Fig. A sketch of a collision between two particles

Before the collission the beam particle moves with four-momentum

$$p_b = (p_{lab}, 0, 0, \sqrt{m_b^2 c^4 + p_{lab}^2 c^2})$$

and the target particle m_t is at rest,

$$p_t = (0, 0, 0, m_t c^2).$$

We first need to determine the velocity v of the Lorentz transformation that bring is to the centre-of-mass frame. We use the Lorentz transformation rules for momenta to find that in a Lorentz frame moving with velocity v along the x-axis relative to the CM frame we have

$$p'_{bx} = \gamma(v)(p_{lab} - vE_{lab}/c^2)$$

$$p'_{tx} = -m_t v\gamma(v).$$

Sine in the CM frame these numbers must be equal in sizebut opposite in sign, we find a linear equation for v, with solution.

$$v = \frac{plab}{m_t + E_{lab}/c^2} \approx c\left(1 - \frac{m_t}{plab}\right)$$

Now if we know the momentum of the beam particle in the CM frame after collision,

$(p_f\cos^{\theta_{CM}}, p_f\sin^{\theta_{CM}}, 0, E'_f)$,

where $^{\theta_{CM}}$ is the CM scattering angle we can use the inverse Lorentz transformation, with velocity - v, to try and find the lab momentum and scattering angle,

$\gamma(v)(p_f\cos^{\theta_{CM}} + vE'_f/c^2) = p_{flab}\cos\theta_{lab}$

$p_f\sin^{\theta_{CM}} = p_{flab}\sin\theta_{lab}$,

from which we conclude

$$\theta_{lab} = \frac{1}{\gamma(v)} \frac{p_f \sin\theta_{CM}}{p_f \cos\theta_{CM} + vE'_f/c^2}$$

Of course in experimental situations we shall often wish to transform from lab to CM frames, which can be done with equal ease.

To understand some of the practical consequences we need to look at the ultra-relativistic limit, where $p_{lab} \gg m/c$. In that case $v \approx c$, and $\gamma(v) \approx (p_{lab}/2m_tc^2)^{1/2}$. This leads to

$$\tan\theta_{lab} \approx \sqrt{\frac{2m_tc^2}{plab}} \frac{\mu \sin\theta c}{\mu \cos\theta c + c}$$

Here u is the velocity of the particle in the CM frame. This function is always strongly peaked in the forward direction unless u c and $\cos\theta_c \approx -1$.

ELASTIC-INELASTIC

We shall often be interested in cases where we transfer both energy and momentum from one particle to another, i.e., we have inelastic collissions where particles change their character - e.g., their rest-mass. If we have, as in Fig., two particles with energy-momentum k_1 and p_q coming in, and two with k_2 and p_2 coming out, We know that since energy and momenta are conserved, that $k_1 + p_1 = k_2 + p_2$, which can be rearranged to give

$$p_2 = p_1 + q, \quad k_2 = k_1 - q.$$

Fig. A sketch of a collision between two particles

and shows energy and momentum getting transferred. This picture will occur quite often.